CANARD
A REVOLUTION IN FLIGHT

COMMEMORATIVE
EDITION

WITH THE ORIGINAL FOREWORD BY
BURT RUTAN

MILITARY - CIVILIAN - HOMEBUILT

ANDY LENNON

An Aeronautical Publishers Book

Published by *Aeronautical Publishers*
AeronauticalPublishers.com

Copyright© 1984 (Original Edition) and
2021 (Commemorative Edition)
by Michael A. Markowski

ISBN 978-0-938716-88-4

On the Cover

The Beechcraft Starship I was a composite construction business aircraft configured to carry two pilots and nine passengers. This 85% scale prototype,
designed by Burt Rutan, enabled engineers to gather pertinent data before
building the full-scale version, startling the aviation world when it was introduced in 1983. Courtesy Beech Aircraft Corporation.

Manufactured in the United States of America

Dedication

This book is dedicated to three young men; brothers Andrew Gilmour, and Peter Llewellyn Price and their cousin Devin Gram Willey. I am proud to be their grandfather.

— A.G. Lennon

Acknowledgement

In supplying data and photographs, the courtesy of the following companies and individuals is gratefully acknowledged. Their contributions add great interest to this book as well as reflecting their fine products.

Rutan Aircraft Factory, Beech Aircraft Corporation, Gates Learjet Corporation, Avtek Corporation, Diehl Aero-Nautical Co., Quickie Aircraft Corporation, CO-Z Development Corp., Air International Magazine, Sun Aerospace Corporation, Goldwing Ltd., Old Man's Aircraft Co., Ultralight Flyer Inc., Merganser Aircraft Corp., Avia Fiber A.G. Flugzeugbau, Viking Aircraft Ltd., Mr. Gian Bezzola, Mr. George B. Collirge, and Mr. Michael A. Markowski.

This acknowledgement would not be complete without reference to the vital contribution that the Experimental Aircraft Association and its members have made and are making to the homebuilt aircraft movement. Without their aid, developments such as Rutan's and many others would not have occurred. EAA's official organ, *Sport Aviation* is one of the very best examples of journalism of its type.

About the Author

Andy Lennon got "hooked" on aviation at age 15 after a short ride in a Curtiss Robin, and later joined the Montreal Flying Club flying D.H. Gipsy Moths and early Aeroncas. He was educated at Edward VII School, Strathcona Academy, Montreal Technical School, McGill University, and the University of Western Ontario.

After a brief stint in banking and life insurance, he entered Canada's infant aircraft manufacturing industry. However, after marrying and starting a family, he opted for more stable employment in general manufacturing, specializing in Industrial Engineering. He moved into management of, among others, a plant producing Sanitary Pottery, of the "Best-Seat-in-the-House" kind, and ultimately to corporate headquarters.

During this period, he continued to be an avid student of things aeronautical, particularly airplane design, studying aviation textbooks, NACA and NASA reports, and reading aviation periodicals. He designed, built, and flew a wide variety of radio-controlled model aircraft that were intended as miniatures of potential full-scale light aircraft.

He continued his full-scale flying activities and was licensed in both Canada and the US, and held memberships in the Canadian Owners and Pilots Association, Experimental Aircraft Association, Montreal R/C Club, and the American Institute of Industrial Engineers. He married a warm and loving lady, raised two wonderful daughters, and lived in a suburb of Montreal.

As a final tribute to a model airplane designer extraordinaire, just before he passed away in 2007, Andy was nominated as a "Leader Member" of the MAAC (Model Aeronautics Association of Canada).

Contents

Foreword

It has been 22 years since my initial experiments with canard aerodynamics. My early inspiration came from two aircraft that were being developed while I was in engineering college. The North American XB-70 and the Saab Viggen had not yet flown, but they appeared to me as beautiful shapes that "looked right" and would probably fly well.

My early work involved applying newly-learned basics to canard types. The emphasis was on trying to achieve natural stall avoidance, rather than on performance optimization. I quickly found that stall avoidance could be achieved, but only after many attempts and refinements in the wind tunnel. I did not have the expertise nor the computer tools to address the performance refinements. That capability came much later. In the meantime, I concentrated on the development of a light aircraft to see if the canard configuration could be successfully applied in that area.

A great deal of my early work was successful only because of some unusually good luck. After I later learned the analysis methods, I found the many traps I nearly fell into during the VariViggen development. I had been lucky indeed. There is a lot to be learned by studying history, but I did not have a summary of past canard design attempts.

Now, Andy Lennon has provided us with an excellent history of the canard and tandem-wing airplanes. We have, for the first time in a single source, descriptions and photos of the tail-first aircraft from the Wright Flyer to the Starship I. The text also details the specific flying qualities exhibited by many of the aircraft.

To the aviation enthusiast and pilot, the book will provide interesting and sometimes fascinating accounts of how the various shapes handled in the air. As you will see when you read this book, the canards have come in a wide variety of shapes and sizes. Some have been well designed and have been able to make use of the benefits of performance improvement and natural stall limiting. Others, though somewhat similar in outward appearance, have suffered from terrible flying qualities and unsafe low

6

speed stability. The individual technical reasons for the failures are beyond the scope of this text.

The designer will find the book useful in highlighting many of the problems inherent in tandem-wing design. A designer with the proper tools generally can avoid a problem area if he is aware of it. The problems that he is not aware of are the ones that provide the most unpleasant surprises. Perhaps the most significant value of Mr. Lennon's presentation will be his success in pointing out the importance of adequate analysis to the amateur-designer. I have lost several personal friends in aircraft accidents because of inadequate concern for complete technical design.

Experience, particularly manned flight test experience, is the best teacher. Items that were deficiencies, such as the power-inducted pitch trim change of the VariViggen or the rain-induced trim change of the Quickie, or the inability to use flaps on most of the configurations, can now be successfully addressed. The best of the most recent designs have shown to be free of any significant compromises.

Tomorrow's canard-type designs will be even better than today's. This is because many designers are not only becoming familiar with the methods to achieve the best performance, but they now have improved computational tools available to optimize the airfoils and to blend the various components. The future aircraft will even offer basic aerodynamic solutions to some old problems that have always plagued the designers, such as the undesirable control characteristics when approaching transonic speeds. I predict the introduction of tandem wing commuters and jumbo airliners by 1995.

So enjoy the book and look forward to an exciting continuation of the Canard Revolution in all areas of aviation.

— *Burt Rutan*

Introduction

Early experimenters found that a wing alone, or a bi-plane wing alone, was longitudinally unstable. Some overcame this instability with aft mounted, non-lifting tail surfaces. The Wright brothers however, went the other way—using lifting surfaces ahead of the main wings. They could be flexed by the pilot to increase or decrease angle of attack for longitudinal control. The canard had been conceived.

This concept was emulated by many of the early designers, but they gradually used a "hybrid" of both canard and rear horizontal surfaces. By 1909, both canard and hybrid were displaced by the rear horizontal tail surface. Even the Wright brothers, in their Model B, abandoned the canard concept. The rear tail has since predominated and has become the "conventional."

The reasons for this change are obscure; a reasonable conjecture is that the daring aerobatic aviators of the day found the rear tailed configuration more agile. It could stall and spin, the canard could not! World War I "dog fighting" certainly favored conventional layouts. It is interesting to note that no canards were designed during the 1914-1918 World War One period.

Another reason for "conventional" layouts predominating, may have been due to the increasing power and weight of engines. When arranged as a pusher, locating the CG in the correct position for a canard was difficult. This lead to the hybrids with their lifting rear tail surfaces that compensated for the rearward CG shift. Finally, putting the engine out in front proved to be the "ultimate" solution.

After World War I, interest in the canard concept continued, but no real breakthroughs occurred. Seldom were more than one or two copies of each design built, until Mignet introduced his homebuilt "Pou-du-Ciel" (Flying Flea). It became quite popular for a time and many were built. However, several fatal crashes occurred. History tells us that it had too narrow a range of stable flight angles. As a result, most governments grounded the mighty "Pou."

Then, in 1975, at the Experimental Aircraft Association's Annual Convention at Oshkosh, Wisconsin, Burt Rutan introduced his "VariEze"

—a homebuilt composite construction canard. The response from EAA members and the aviation community was incredible. The breakthrough had arrived.

Rutan, almost single-handedly, has created a pivotal point in aviation history. More versions of his "VariEze" and "Long-EZ" have been built than any other single homebuilt aircraft design. The canard concept had finally caught on from ultralights and gliders to transports and fighters.

General Aviation has entered the "Canard Age." In October 1983, Beech unveiled their "Starship 1," a beautiful design, obviously reflecting Rutan's influence. The Old Man's Aircraft Company is flying their OMAC 1, readying it for FAA Certification. Avtek is working on their "Avtek 400" of composite construction, due to fly shortly. And, Gates-Piaggio introduced their "GP-180"—a hybrid—with both canard and conventional tail-last surfaces. History has certainly repeated itself.

All four of these canards are turboprop powered pushers. The light weight of these powerful, aft-mounted engines greatly simplifies their designers' problem in locating the center of gravity properly. It's not the problem it was with earlier canards.

Perhaps we can say that, the Wright brothers were "right" all along!

Section I

Canard
Stall-Free Safety

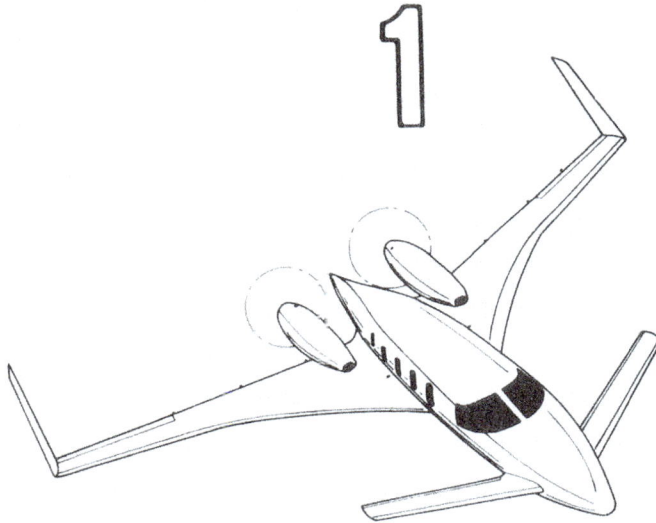

Conventional Aircraft Stall

Basically, a wing stalls when the lift it is producing becomes less than the weight it must support. Wing lift varies with the square of the airspeed. Reducing airspeed by half, reduces lift to one quarter. However, wing lift varies directly with angle of attack, up to the angle of stall, after which it drops off very quickly. Doubling the angle of attack below the stall doubles the lift.

Low speed level flight is achieved with the wing held at a high angle of attack, below the stall. High speed level flight is obtained at low angles of attack. A range of speed from just above the stall to the maximum airspeed is available in between.

A student pilot practicing stalls at altitude, in level flight, slows his plane down by throttling the engine to idle while slowly adding up-elevator. As the angle of attack increases, the airflow over his wings deteriorates from a smooth flow, to a condition where it starts to "separate" at the wing's upper surface trailing edge. The wing starts to "mush" downward, and then the airflow separates completely. The wing is fully stalled. At altitude, this is a safe maneuver, but close to the ground it is very dangerous and often fatal.

In level flight an airplane has only to support its weight, including fuel, cargo, pilot and passengers. In turning flight however, centrifugal force causes a substantial increase in the weight the wing must support. This happens when an aircraft banks, is pulled up sharply, or comes out of a dive. During a loop, centrifugal force is sufficient to hold the pilot in his seat when he is upside down.

For example: a Cessna 172 at gross weight in level flight, will stall at 57 mph. In a 60° banked turn, the stall speed increases to 81 mph, 42% higher, due to the extra load imposed by centrifugal force. In slow speed flight a moderate increase in bank angle raises stalling speed above straight and level stall speed, and a stall results.

There is another factor—wing pitching moment. It depends on the airspeed, wing section characteristics, wing chord and area, and, in level flight, tends to cause the aircraft to nose down. It is controlled by the tailplane—either behind in "conventional" aircraft, or ahead in a canard. At the stall, this nose down pitching

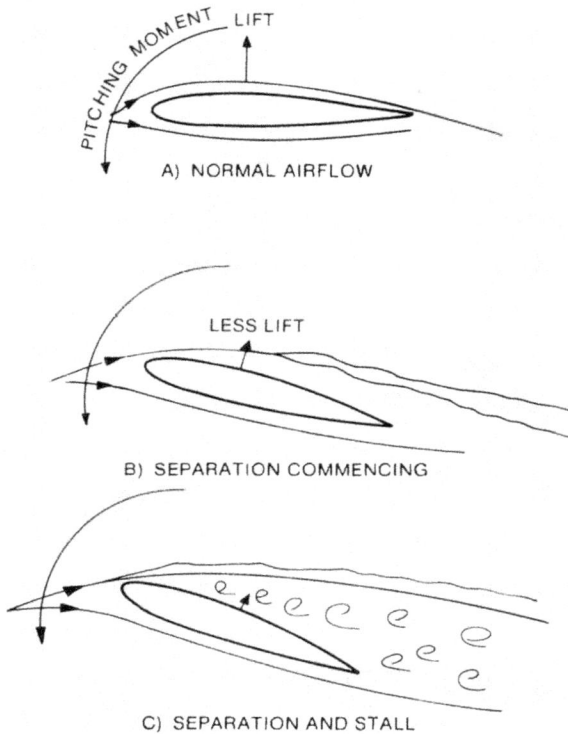

Fig. 1-1. The mechanics of the stall.

moment increases. This is stabilizing, since it tends to reduce the angle of attack. However, combined with wing stall, it is dangerous.

Conventional Aircraft Stalls

A conventional aircraft has its center of gravity ahead of the center of lift (neutral point). The nose down effect of both weight and pitching moment is overcome by the aft horizontal tailplane, achieving balance. The wing must then support not only the aircraft's gross weight, but also the balancing down load from the horizontal tail.

After the wing stalls, and the aircraft pitches down, it dives until speed builds up, and normal airflow is established, allowing the wing to lift again. Judicious use of elevators and power can reduce the extent of this recovery dive.

This is all very well, provided both left and right portions of the wing stall evenly and simultane-

ously. If, however, one panel stalls first (due to a gust, sharp aileron action or deliberate use of full rudder) a "spin" results. In a fully developed spin, an aircraft rotates around a vertical axis—the inside wing panel is fully stalled while the opposite panel is still lifting. Recovery can only be achieved by use of down-elevator to unstall the inside wing panel, plus full rudder opposite the direction of rotation. A spin is exciting at altitude, but close to the ground it is usually fatal.

Aircraft designers have recognized the stall/spin syndrome and modern lightplanes incorporate features to reduce the incidence of such accidents. Here's a few: wing twist, more cambered wing tip sections (Ref. 6), fixed slots ahead of the ailerons (Ref. 8), leading edge droop (Ref. 10), differential aileron movement (the down going aileron moves much less than the up going (Ref.'s 28-29), spoilers replacing ailerons (Ref. 31), improved cockpit visibility (Ref. 39), and last but not least, the stall warning horn.

The horn is an electro-mechanical device. It's composed of a hinged, spring-loaded blade placed in a horizontal position in the wing leading edge. An electrical circuit connects the blade to a speaker in the cockpit. As the wing approaches the stall, air pressure on the blade compresses the spring, making electrical contact, actuating the horn. This device is effective, but dirt or bugs can prevent its operation.

Wing flaps when deployed, reduce the stall speed but lower the stall angle of attack by two or three degrees.

Look at Fig. 1-3, the "Stall Crash Scenario." While this is imaginary, it vividly portrays the danger inherent in conventional aircraft.

A young, inexperienced pilot is making an approach to landing, and is on the down wind leg. Note the wind direction. As he turns onto base, he throttles back, reduces speed, lowers his flaps and starts his turn to final. Since he is flying a high wing aircraft, the left wing obscures his view of the runway until he reaches point A. At this point, his airspeed is 70 mph—but he realizes that the wind has drifted him well to the right of the runway. He cranks in left aileron and up elevator to tighten his turn in order to line up with the runway. At

PITCHING MOMENT

LIFT

WEIGHT

BALANCING TAIL DOWN LIFT

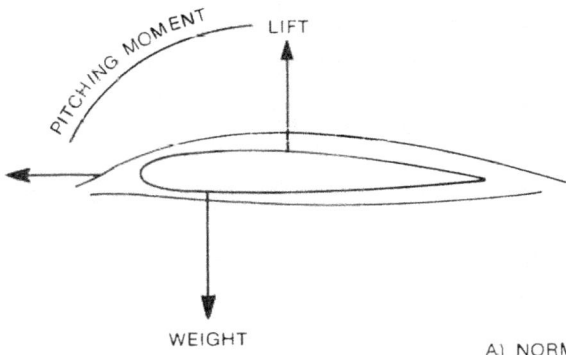

A) NORMAL FLIGHT

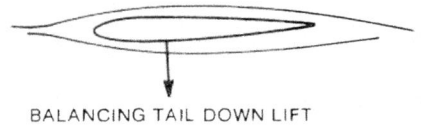

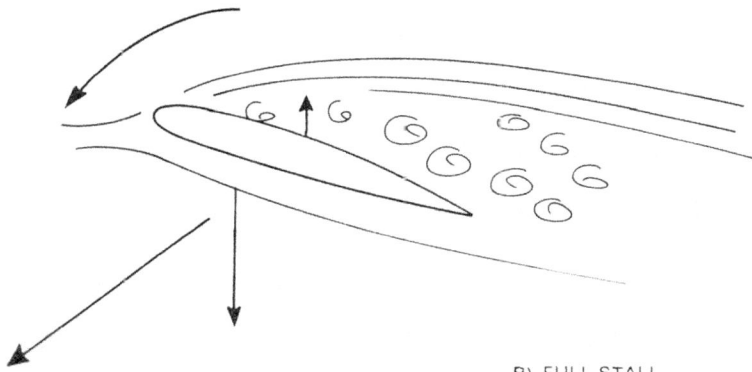

B) FULL STALL

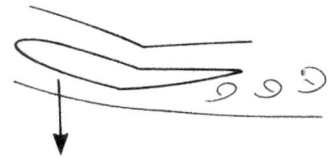

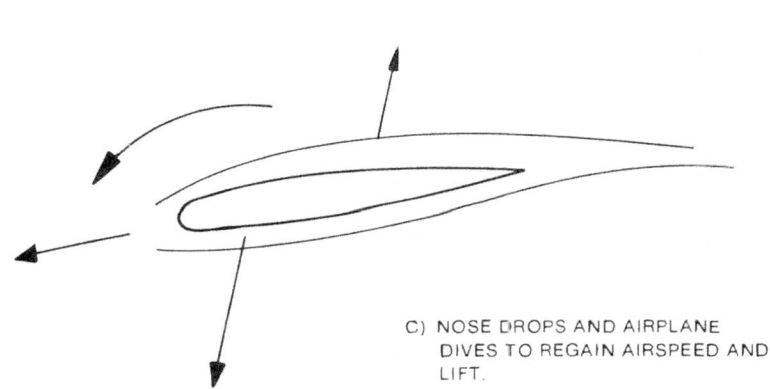

C) NOSE DROPS AND AIRPLANE
DIVES TO REGAIN AIRSPEED AND
LIFT.

Fig. 1-2. The "conventional" aft-tail aircraft stall.

13

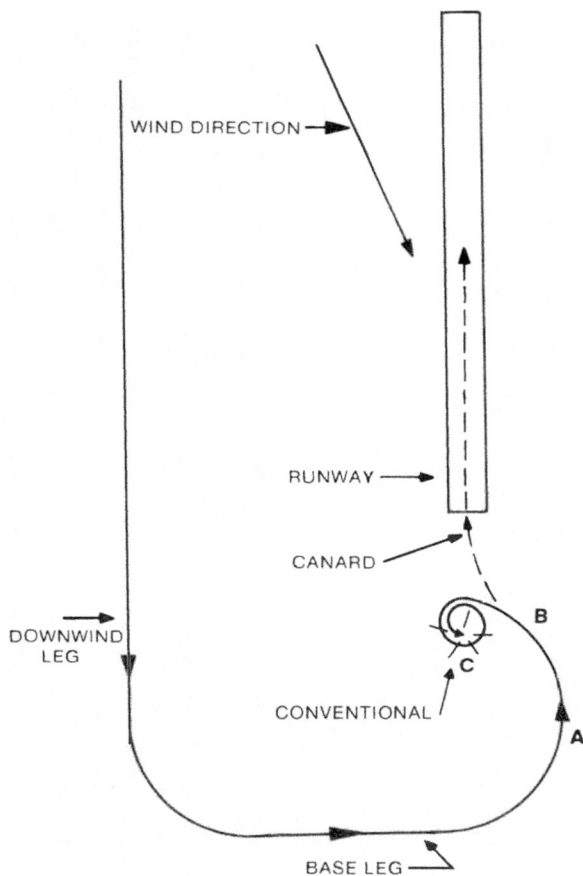

WIND DIRECTION →

RUNWAY →

CANARD

DOWNWIND
LEG

B

C

CONVENTIONAL

A

BASE LEG

Fig. 1-3. The stall/spin-crash scenario for a "conventional" aircraft.

point B, the stall warning horn blares. Nervously, he applies full opposite aileron and a touch of up elevator to hold his nose up. The left wing stalls at 300 feet above ground level; over he goes, crashing at "C."

What he *should* have done at point B was give down elevator and apply power BEFORE reducing his bank, this would have prevented the wing from stalling.

The author experienced just this sort of circumstance in a Piper "Cherokee" during a landing approach to Sanford Municipal Airport in Maine. Fortunately, the correct recovery procedure was used, or you would not be reading this.

So what happened to our forever-young pilot? Sharply increasing his bank at point A, he greatly increased the centrifugal force load on his wings, to a level beyond the wing lift capability at his airspeed. The wing mushed; down left aileron increased the wing angle of attack, induced the left wing to stall; and the aircraft rolled to the left into a spin and crash.

Statistically, stalls are the highest single factor causing fatalities, usually occurring while the aircraft is flying slow at low altitude, during takeoff and landing. The canard, with its stall free aft wing, is much safer, accounting for its increasing popularity.

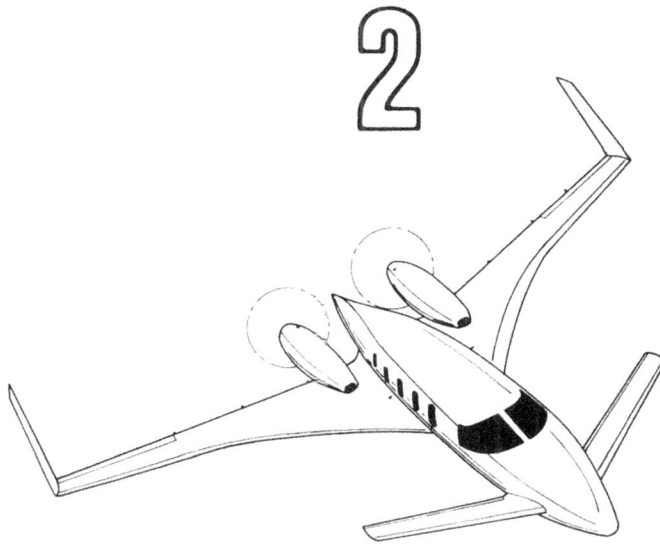

Canard Aircraft Stall

The canard airplane has two lifting surfaces, with the forward wing lifting a greater share of the total weight per square foot of wing area (i.e. it has a heavier wing loading) than the aft wing. This is achieved by having the center of gravity well ahead of the aft wing. The aft wing pitching moment also adds to the foreplane load.

In a well designed canard, the forward wing will always stall at a lower angle of attack than the aft wing. The latter cannot be accidentally stalled, so that the ailerons remain effective. Centrifugal force, in turning, climbing or diving flight only hastens the foreplane stall, which then mushes downward, as shown in figure 1-4, reducing its angle of attack, resuming its lift. The aft plane remains unstalled; lateral control is maintained and spins are not a problem. Actually, the canard can be flown with the foreplane alternately stalling and unstalling, the nose bobbing up and down gently in a porpoising mode.

A canard can be made to spin—or more correctly forced into a tight spiral. A canard will not voluntarily spin like a conventional rear tail airplane.

Recovery from a tight spiral is prompt.

Flaps for low speed landings present a design problem in canards. Foreplane maximum lift capability, even though it is augmented by the elevator, limits the extent of aft wing flap area and angle of deployment.

Too much lift aft of the CG will cause a nose down action. Also, since the foreplane stalls before the aft plane, the latter cannot achieve its maximum lift potential. However, the flapped foreplane lift offsets this factor.

The hybrid aircraft, with both foreplane and aft horizontal tail surfaces, overcomes this problem. Elevators on the tail overcome lift imbalance between fore and aft wings—permitting larger flaps and lower landing speeds, but at the cost of more weight and drag.

Returning to Figure 1-3, the "Stall Crash Scenario"—and with our young pilot flying a canard aircraft, he arrives at point A; steepens his angle of bank to sharpen his turn and the foreplane stalls. His aircraft noses down, but since his

15

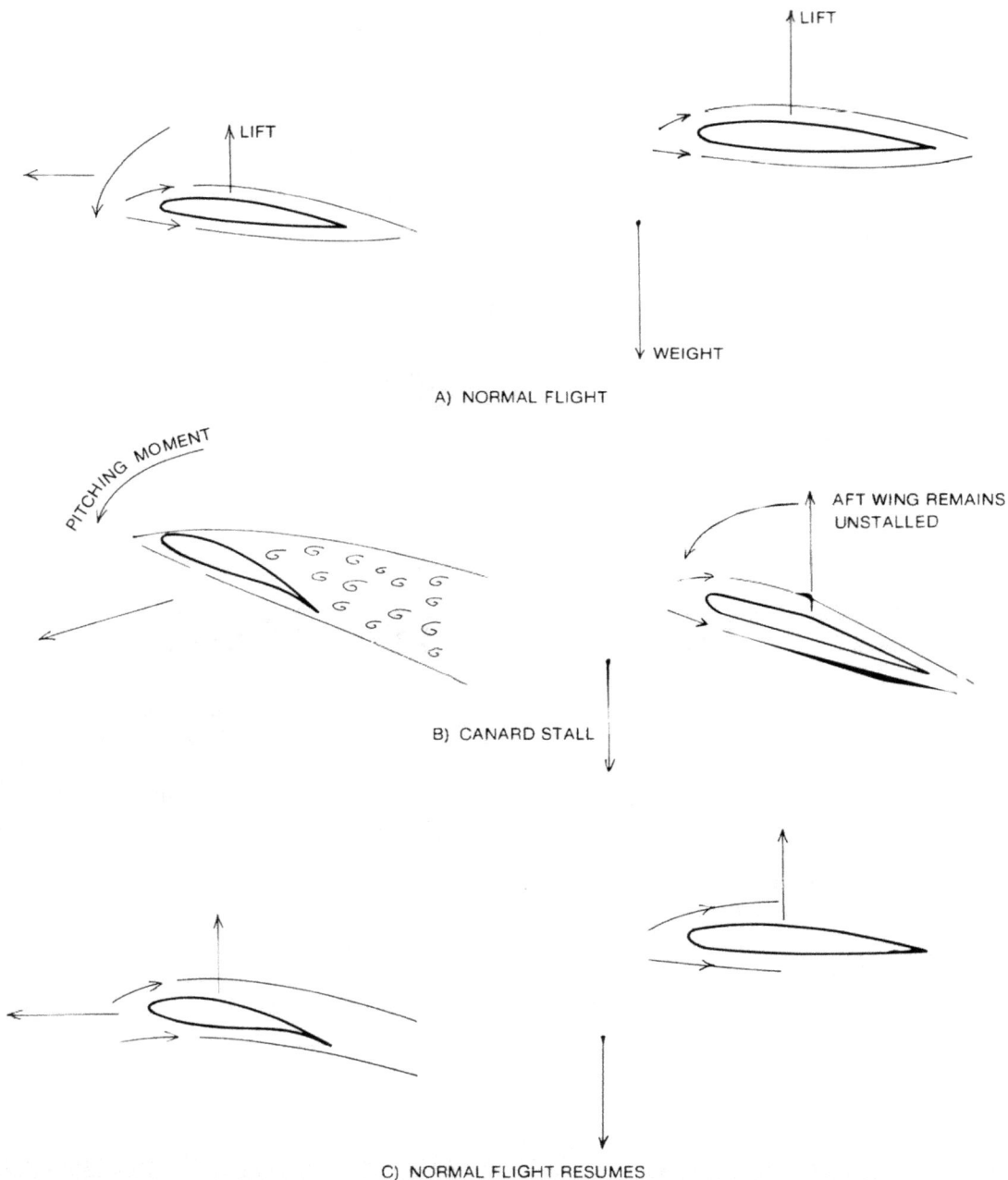

A) NORMAL FLIGHT

B) CANARD STALL

AFT WING REMAINS
UNSTALLED

C) NORMAL FLIGHT RESUMES

Fig. 1-4. The canard aircraft "stall."

main wing is unstalled, he can safely reduce his angle of bank with ailerons—he corrects the flight path and continues to a normal cross wind landing.

This is the major safety feature of the canard.

Admittedly it is achieved at the expense of limited aerobatic capability, but the airplane is becoming, more and more, a means of fast transportation—aerobatics are of no concern to safe flying.

The canard is the shape of the future.

Section II

Experimental Aircraft
And The
Development
Of The Canard

3

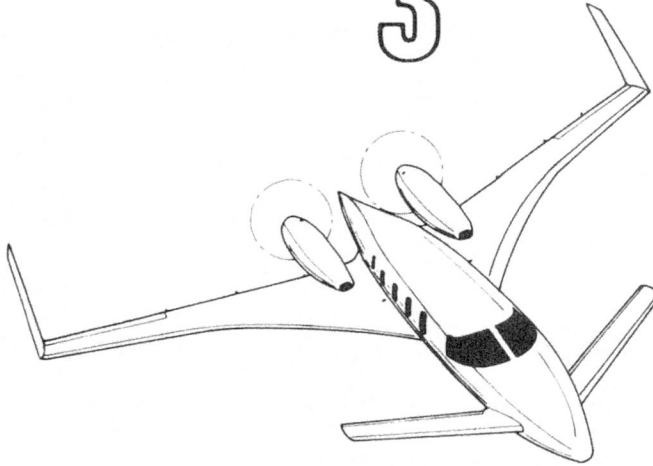

Canard And Tandem Wing Aircraft

The Wright Brothers—USA 1900-1909

Wilbur and Orville Wright were sons of Bishop Milton Wright of the United Brethern Church, and Susan Kolmer Wright. They had two older brothers and a sister.

The two had an unusually close relationship. Everything they accomplished was the result of conversations, suggestions and downright arguments. They worked together as a team. However, neither finally graduated from high school nor attended college, and remained bachelors all their lives.

In 1892, the brothers formed the Wright Cycle Company to sell, repair, and, ultimately manufacture, their own brand of bicycles. This business provided the funds for their work in aviation. They were never financed by outside interests.

A famous German aviation pioneer, Otto Lilienthal, was an inspiration to the Wright brothers. His writings, based on his hang gliding experiments, were invaluable to the Wrights and became the basis for their own experiments. Lilienthal stalled, crashed and died in a glider

accident in 1896, but this did not deter the Wrights from proceeding with their study of aeronautics—it made them more cautious. They also became familiar with the work of Langley and Chanute, and in 1899, they began experimenting.

Early on, they realized that control of a flying machine by shifting the pilot's body to restore equilibrium, was not only unsatisfactory but unsafe—it had lead to Lilienthal's death and that of England's early pioneer, Percy S. Pilcher. They were the first to conceive of lateral control by wing warping—today we use ailerons.

In 1899, they proceeded to experiment with gliders of five foot span flown as kites at Dayton, Ohio, with the wing warping controlled by strings from the ground. Using this same glider, they also provided for longitudinal control after testing horizontal surfaces both ahead of and behind the wing. As the world knows, they settled on the canard surfaces in front of the aircraft.

At this point, the brothers believed they had solved the problems of both lateral and longitudinal control. However, in subsequent glider trials they discovered the absolute necessity for directional control...the hard way.

Kitty Hawk

The brothers now proposed to build a full size, man-carrying glider on which to prove their control methods. They were daring, yet not foolhardy. They fully realized the risks of flight, yet were determined to minimize these risks by good design, and sturdy construction. They deliberately flew in high winds to reduce ground speed and ground impact, preferably over sand.

They needed flat open areas with steady winds as well as small hills, and freedom from shrubs or trees for their projected low level flights. Dayton and its surroundings were unsatisfactory. After consulting with the U.S. Department of Agriculture Weather Bureau and others, they selected Kitty Hawk, North Carolina as the site for their experiments.

In early 1900, the Wrights began assembly of components and material for a full size man-carrying glider to test both their wing warping and longitudinal control methods. In September 1900 Wilbur journeyed to Kitty Hawk—a long, tedious and tiring voyage at the turn of the century. Orville subsequently joined him, and together they set up a tent and began preparations for their tests.

The first glider was a biplane. It spanned 17 feet, had a total wing area of 165 square feet and, with pilot, weighed 190 pounds. Wing loading was 1.15 pounds per square foot. The machine had no rudder.

The horizontal control surface was placed ahead of the wings for longitudinal stability, and it was under angle of attack control by the pilot. The Wrights believed the forward position offered better control and one which would avoid the nose dives similar to those that killed Lilienthal and Pilcher.

At first the glider was flown as a kite, but the brothers were determined to try piloted gliders. They journeyed to Kill Devil Hill, a sand dune about 100 feet high, 4 miles south of Kitty Hawk. Takeoffs were made by running downhill into the wind. When lift was apparent, the pilot jumped

aboard and glided free down the slope of the hill for 300 to 400 feet, at an altitude of 3 to 4 feet above the sloping terrain. Landings were made on skids in soft sand, without damage to the glider or injury to the brothers.

The pair found their lateral and longitudinal control methods were effective, but their glider had inadequate lifting power. In their methodical way, they measured lift and drag versus angle of attack, concluding their wing's curvature, or camber, was inadequate. They were beginning to feel that Lilienthal's tables of lift were inaccurate.

It should be noted at this point that their glider had no dihedral—the wing tips were lower than the wing center section i.e., anhedraled (or cathedraled). Having tried dihedral, they concluded that it posed serious problems of controls in the turbulent winds in which they flew. This carried through into their first powered flights, but later models had straight wings—no dihedral or anhedral.

The brothers firmly believed that further trials were necessary in order to learn how to design wings properly and to improve on their control methods. Then, and only then would they be in a position to contemplate use of power for sustained flight.

Glider Experiments 1901

This year the brothers chose to set up their tent close to Kill Devil Hill. They also erected a rough frame shed to house their new glider, and be used as a workshop.

The new glider had a wingspan of 22 feet, a total lifting area of 290 square feet and a total weight with pilot of 242 pounds—wing loading was 1.2 pounds per square foot of wing area, slightly higher than their 1900 machine. The wing camber was set at 1 in 12 (i.e. 8½%), as prescribed by Lilienthal. To their dismay this glider proved to be much more difficult to control longitudinally and would "stall." Happily however, the canard arrangement prevented any serious accidents. The canard is really what "saved" the Wrights and allowed their eventual success.

They reduced the wing camber to 1 in 18 (i.e. 4½%). This improved their longitudinal control to that of the previous year. However, they found that the wing warping did not always produce the desired result. They encountered what is now known as "adverse yaw" where the down-going wing trailing edge, in addition to providing added lift, also created considerably more drag. This caused the glider to yaw away from the desired direction of the turn. The Wrights now became convinced that something more was needed to control their gliders.

They concluded that the airfoil theories of the day were unreliable. It was apparent that more accurate data was needed before the problems of flight could be solved. The Wright brothers were understandably discouraged, and all but discontinued their work. With the encouragement of family and friends however, they continued.

Wing Tunnel Tests 1901

In 1901, aerodynamics was not even in its infancy—it was still embryonic. The brothers decided to start from scratch, discarding all previous data. They would develop their own aerodynamic data. They built a wind tunnel, crude by todays standards, but none the less effective. It consisted of a box 6 feet long by 16 inches square inside. A flat bladed fan forced a current of air through a funnel shaped entry equipped with a honeycomb to reduce the turbulence created by the fan. Two ingenious balances were made to measure lift, drag and angles of attack.

Fig. 2-1. The Wright brothers used their 1902 glider to perfect their flying skills, and prove the effectiveness of their control system. The canard configuration "saved" them from disaster many times, by inherently avoiding the dreaded stall. Courtesy Smithsonian Institution.

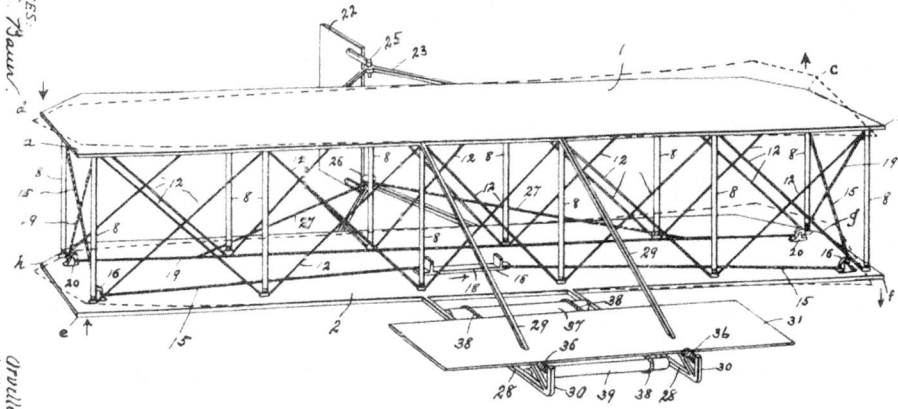

Fig. 2-2. Wrights' patent drawings reveal the secrets of their canard and three axis control system—the heart of all successful flying machines.

In two months near the end of 1901, 200 surfaces were tested, including monoplanes, biplanes and triplanes. They tested the effect of aspect ratio, camber and airfoil shape on lift/drag ratios.

This work was carried out in their usual methodical, industrious and quiet way, and yielded a data-bank on the aerodynamic properties of wings, control surfaces, and structures. It exceeded anything that was then available or was to be available for another 10 years. This information was probably the Wright brothers greatest

contribution to aerodynamic knowledge, and formed the foundation for later scientific testing carried out world wide by such organizations as NACA —(National Advisory Committee for Aeronautics) since superceded by NASA (National Aeronautics and Space Administration).

Glider Experiments 1902

By the end of August 1902, the brothers were back at Kill Devil Hill with a new glider developed from their wind tunnel testing. This machine had a

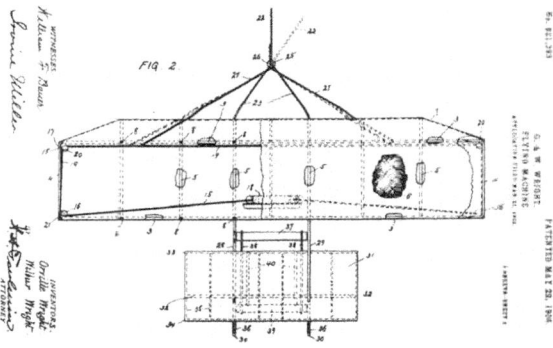

Fig. 2-2. (continued).

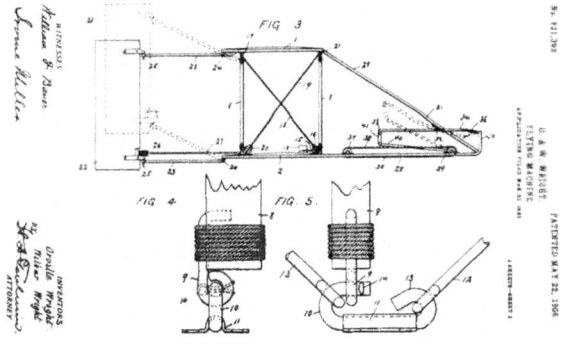

Fig. 2-2. (continued).

wingspan of 32 feet 1 inch, an area of 305 square feet. Its gross weight was 256 pounds, yielding a wing loading of 0.83 pounds per square foot, which was lower than the previous gliders.

The biplane structure of the previous machines was retained, since it provided the lightest, strongest structure for its area—yet it could be warped readily for lateral control. An aspect ratio of six was employed, based on the wind tunnel findings.

This was the first glider to have fixed twin vertical fins behind the aircraft, designed to overcome the adverse yaw discovered the previous year. The fins presented a new problem, however. When a gust lifted a wing and caused a side slip towards the low wing, wing warping opposed to the side slip caused an increase in drag of the low wing that led to a spiral dive. The vertical aft surfaces actually contributing to the spiral by resisting the corrective sideways motion. The result was worse than that experienced with no vertical tail.

The brothers then conceived the idea of a single, moveable rudder that would be arranged to move in concert with the wing warping. They believed the rudder side force would oppose the added drag of the wing tip at the higher angles of attack. This did two things. In addition to offsetting adverse yaw it tightened the aircraft's turn and permitted centrifugal force to oppose any side slip. When opposite warp was used to recover from a sharp gust induced bank, the rudder opposed the adverse yaw.

With the addition of the rudder, the solution to the problem of three axis control had been achieved. The Wrights felt they could now concentrate on the problem of powered flight.

One of the benefits of their canard design again became evident during the 1902 Trials. Whenever

Fig. 2-3. That incredible first flight—Orville Wright lifts the Wright Flyer off the sands of Kitty Hawk, North Carolina on December 17, 1903. The world's first successful airplane—a canard! Courtesy Smithsonian Institution.

WRIGHT AEROPLANE, 1903
NOMENCLATURE & DETAILS

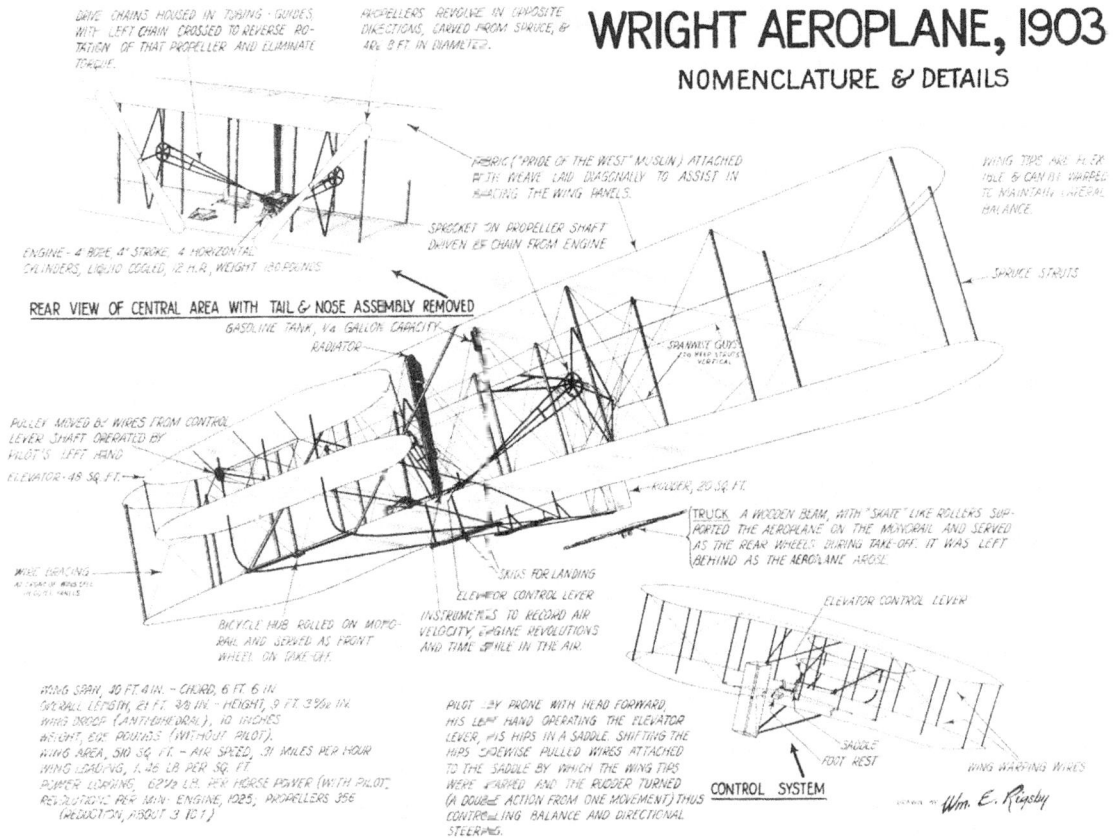

REAR VIEW OF CENTRAL AREA WITH TAIL & NOSE ASSEMBLY REMOVED

CONTROL SYSTEM

Drawn by Wm. E. Rigsby

Fig. 2-4. Details and nomenclature of the Wright brothers' Kitty Hawk Flyer. Courtesy Smithsonian Institution.

the aircraft nosed-up sharply, the elevator stalled, while the wing remained unstalled. The glider safely "mushed" downward instead of falling off on a wing. It did no spin into the ground as aft tail gliders were apt to do. On several occasions, this feature permitted the Wrights to avoid fatal accidents of the type suffered by Lilienthal and Pilcher.

The Engine And Propeller

Unable to find any existing gasoline engines that met their specifications of 8 to 9 horsepower, while weighing 180 pounds and relatively vibration free, the Wrights decided to design and build their own. With the help of their mechanic, Charles E. Taylor, they succeeded.

The new engine was shop tested in May 1903. It had 4 horizontal cylinders in-line (with bore and stroke of 4 inches each), and an aluminum alloy crankcase and water jacket. Developing just under 12 horsepower it weighed, complete with radiator, a little more than 200 pounds. This unit powered the first true flights.

The Wrights assumed that propeller design would not be a problem, since marine propeller technology had existed for years. They were wrong. Once again, they had to develop their own theory. They had neither the time nor the funds that extended "cut and try" experimentation demanded. In their methodical, painstaking way, they proceeded to develop their own formulas and were able to predict their performance. Subsequent tests proved their calculations of thrust correct to just under 1% less than predicted.

Powered Flight 1903

The 1903 powered airplane had a wingspan of 40 feet, 4 inches; a camber of 1 in 20, and a wing area of 510 square feet. With the pilot aboard, it weighed 749 pounds and had the highest wing loading so far; 1.47 pounds per square foot. The plane was not symetrical. The right wing was 4 inches longer than the left, to compensate for the engine's weight at right of center, being greater than the pilot weight at left of center. To reduce drag, the pilot lay prone, as with the gliders.

The same elevator and wing warping controls were employed. However, instead of a single rudder, twin moveable rudders were installed to the rear. They were linked to the wing warping mechanism and worked simultaneously with it.

To neutralize gyroscopic effects, twin counter rotating propellers were installed in pusher configuration, driven by chains running over sprockets. Crossing one chain in a figure eight achieved the counter rotation. Propellers were eight feet 6 inches in diameter.

Practice flights with the 1902 gliders, bad weather, and problems with the propeller shafts and sprockets, delayed the first powered flight until December 14, 1903. Assisted by five local lifesavers, the machine was placed on a dolly. It ran on a metal surfaced sectional track which provided "wheels" temporarily, until the aircraft became airborne down the Kill Devil Hill slope. Wilbur, winning a coin toss, piloted the machine. He took off after a run of 35 to 40 feet, but too much up-elevator caused a stall and the typical canard mush to a landing after only 3½ seconds in the air. Since the plane landed at a lower altitude than its takeoff, the flight was considered unsuccessful. Minor damage was repaired.

December 17, 1903 - The First Powered Flight

December 17 was a cold, windy day, but the brothers were anxious to proceed. Selecting a level site, the track was laid out facing into the wind. The airplane was mounted on its dolly which in turn rode on the single track.

This time, Orville was the pilot. At 9:35 A.M. the machine took off after a run of forty feet. It climbed to 10 feet, and flew 120 feet in 12 seconds. Because of winds and too sensitive an elevator,

the flight was eratic. In the fourth flight at noon, Wilbur flew 852 feet in 59 seconds, despite the same problems of gusts and sensitive elevator control. Shortly thereafter, the aircraft was tumbled by a heavy gust of wind, and damaged beyond repair. It never flew again, but the Wright brothers had proven that powered flight was possible.

Due to the almost secretive manner in which the brothers conducted their experiments, it was some years before the world would realize their significance. The Wrights continued with their trials undaunted.

The Wright 1904 Machine

Encouraged by their success, the Wright brothers continued. An improved airplane, Flyer No. 2, was flown in 1904 and resulted in better control and maneuverability, but was far from being an outstanding success. Attempts to fly the machine in Dayton, before the press, resulted in dismal failure and no small amount of adverse criticism. After some study and thought, the brothers realized that they needed assistance accelerating the machine for takeoffs.

They devised a catapult—a heavy weight raised high in a small tower with a system of cables, running below the track to the far end through a pulley and back to the Flyer. When the weight was released the machine was pulled forward into flight. The weight, originally 800 pounds and finally 1600 pounds, was pulled to the top of the tower by a line of people hauling on the rope. This catapult arrangement made the brothers independent of wind for assisting their takeoffs.

Further flights were made in 1904. The longest, on December 1, lasted five minutes and eight seconds and covered about three miles.

The Wright 1905 Machine

The 1905 machine was longer and had a more powerful engine of 25 horsepower, and a wing area of 503 square feet. Weight with pilot was 854 pounds, putting the wing loading at 1.70 pounds per square foot—the highest ever tried.

The most important change on this model was the separation of the joint wing warping and rudder controls. They now had individual controls for each of the 3 axis, pitch, roll and yaw. The

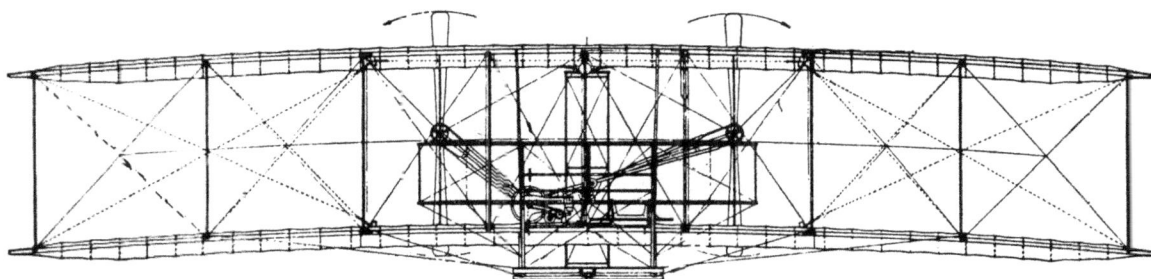

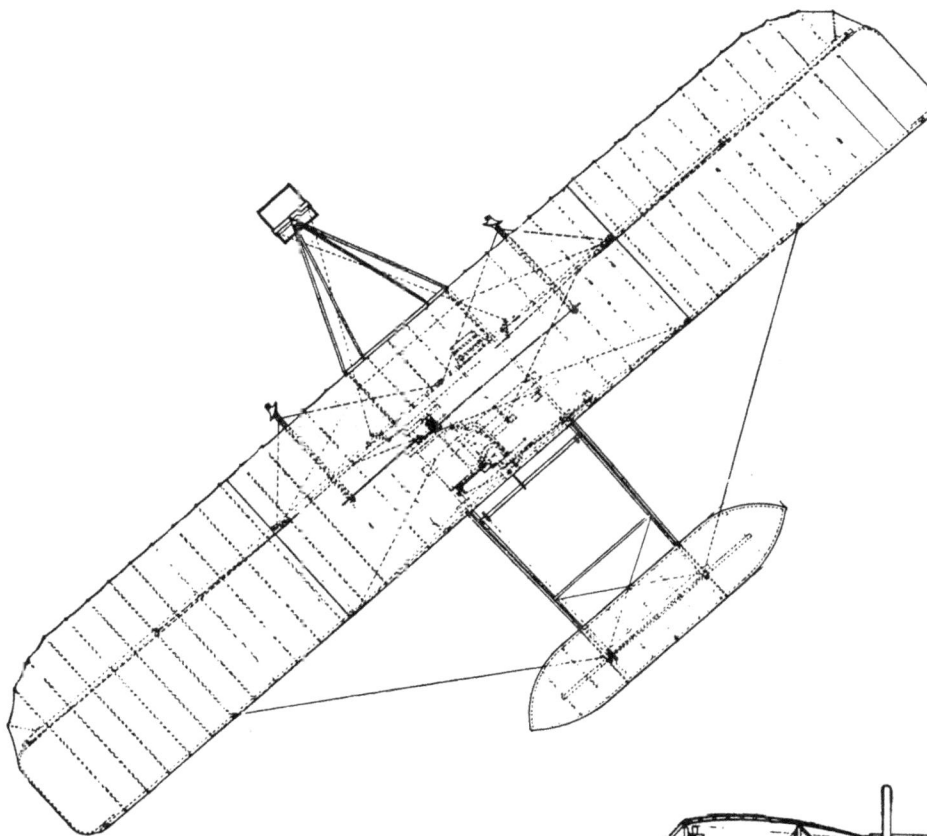

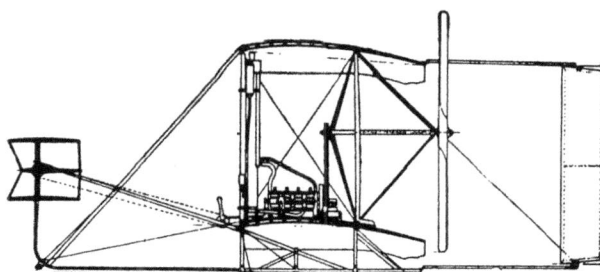

Fig. 2-5. Three-view drawing of the 1903 Kitty Hawk Flyer.

Fig. 2-6. This shot of the Wrights' 1905 Flyer shows its canard to good advantage. At this point, the brothers were still flying the aircraft in a prone position.

control system used on all aircraft today, was complete.

On October 5, 1905 Wilbur Wright made the longest flight of 39 minutes and 28 seconds. He had flown 24.2 miles and circled the field 29 times.

For improved stability and control, the machine was lengthened from 21 feet for the 1903 Flyer to 28 feet in 1905. Subsequently, an even more powerful engine of 35 horsepower was installed. It became evident to the Wrights that the prone position was impractical and very tiring. Provision was made for upright seating, not only for the pilot but also for a passenger.

This machine was offered to the U.S. Government but was refused by the Army, without making any investigation into the stage of development the Wrights had achieved. British and

French interests also came forward, but, until 1908, nothing transpired.

In February 1908, the United States War Department finally contracted with the brothers for an airplane. Shortly thereafter, the Wrights formed a syndicate to license, manufacture, and sell the Wright airplane in France. On August 8, 1908, Wilbur demonstrated his airplane at Le Mans, France, taking off, circling and performing a perfect landing before a highly enthusiastic crowd.

Some time later, Wilbur had a lady for a passenger but insisted she tie a cord around her long ankle length skirts, for obvious reasons. It developed a new fashion, hobble skirts, which were very popular for some time thereafter.

Orville's first public flight took place on Sept-

Fig.2-7. The Wrights' 1909 Signal Corps machine—the world's first military aircraft.

ember 3, 1908 at Fort Myer. However, his final flight ended in a crash which seriously injured Orville and killed his passenger Lt. Thomas Selfridge.

In 1909, Orville completed the government's test flights by flying 10 miles in 14 minutes or about 43 mph. The United States Army formally accepted its first airplane from the Wrights on August 2, 1909.

Seven airplanes were built by the brothers during the period from 1907 to 1909. Commercial companies were found in France and Germany to manufacture Wright airplanes. The Wright Company was organized in the United States with Wilbur as president and Orville as vice-president. Shrewd businessmen, the pair grew wealthy as well as famous.

Tragically, on May 30, 1912 at age 43, Wilbur Wright died of typhoid fever. Orville survived him by 36 years. Their 1903 flyer is now on display at the National Air and Space Museum, Washington, D.C. In 1932 the Wright Memorial shaft in gray granite, 60 feet high, was erected atop Kill Devil Hill, at Kitty Hawk, N.C. to commemorate the great contributions of Wilbur and Orville Wright to aviation.

As a matter of interest, after Wilbur's death, Orville became interested in water flying, and developed the "Model G Aerobat." This was a "conventional" aft tail configuration, with a single step central hull and wing tip floats for lateral, on-the-water, stability. The engine was a 60 horse-power 6 cylinder Wright, located in the forward hull. Typical Wright twin pusher propellers, aft of the biplane wings, were driven by shafting and gearing in the hull.

The Aeroboat weighed 1200 pounds empty and carried a useful load of 600 pounds; one third of its gross weight—an excellent ratio for its time.

27

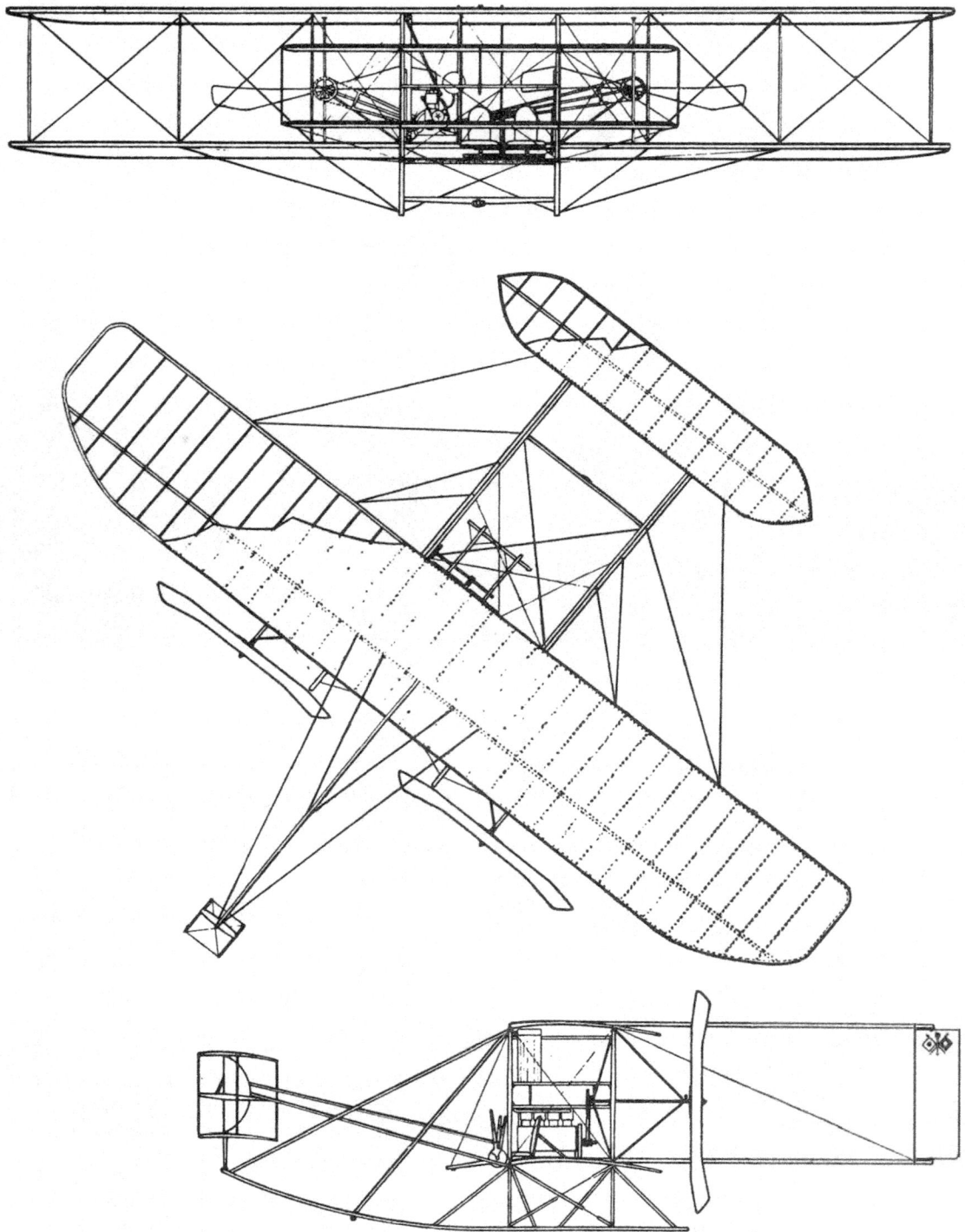

Fig. 2-8. Three-view drawing of the Wrights' 1909 Military Flyer.

Fig. 2-9. Santos Dumont's 14 bis was the first aircraft to fly in Europe in 1906. Canard "box" was used for both pitch and yaw.

OTHER CANARD OR TANDEM WING AIRPLANES AFTER THE WRIGHTS

Santos Dumont 14 Bis-France 1906

Alberto Santos-Dumont, born July 20, 1873, was the son of a wealthy Brazilian coffee planter. He travelled extensively in Latin America and Europe, and was fluently multilingual.

In 1897 he took up residence in Paris, France where he began the design and construction of dirigible balloons. He built twelve over a period of seven years. He won international fame by piloting his No. 6 airship from the French Aero Club grounds at St. Cloud, to the Eiffel Tower and back in less than 30 minutes. This won him the Deutsch Prize of 100,000 francs.

News of the Wright brothers successful flights had caught complacent Europe by surprise. However, it was not until Wilbur Wright's successful flights in August 1908 at Le Mans, France that Europeans learned the true dimensions of the Wright brothers accomplishments.

In 1904, Santos Dumont turned his attention to heavier-than-air machines, encouraged by prizes from the Aero Club of France, in an effort to recapture France's lead in aeronautics. After several attempts at gliders and helicopters, he developed the 14 bis in 1906.

In August and September 1906, at the Bois de Boulogne, a Paris municipality, the aircraft took off under its own power for a hop of four to seven meters. In October, he accomplished flights of 50 meters at an altitude of three to four meters. For this he won the Coupe Ernest Archdeacon. He was widely acclaimed for this feat, but was unaware that the Wrights had made flights of up to 24 miles in 1905. In November 1906, he flew 222 meters (720 feet), in 22 1/5 seconds. While attempting to avoid some over enthusiastic onlookers, he tried to veer, but found his machine was incapable of a small turn. His machine slowed, the canard feature took over and he landed safely

The 14 bis was a biplane canard aircraft—practically a tail first box kite. The interplane struts on both fore and aft wings were fabric covered. Those on the aft wing served as vertical fins for directional stability. While those on the forward wing acted as rudders. In addition to being controlled incidence-wise, the foreplanes could be moved laterally to provide rudder action, act unlike the front steerable wheels of an automobile.

29

The aft biplane wings were heavily dihedralled at ten degrees, but the early model had no wing warping or ailerons for lateral control. Subsequent changes, of which there were many, introduced a form of aileron between the aft wings in the outer bay. These probably combined the functions of lift change and added drag to overcome adverse yaw. Since the pilot's hands were busy with the pitch and rudder controls of the foreplane, aileron control was achieved by lateral motion of the pilot's body through use of a special coat to which the aileron control cables were attached.

The fuselage was a fabric covered box structure. The pilot stood in a large wicker basket just ahead of the aft wings. Behind him was the engine, driving a six foot diameter aluminum propeller of his own design. An extension shaft located the propeller some distance behind the aft wing's trailing edges. Engines were eight cylinder Antoinettes. Santos started with a 24 horsepower version which was replaced by 35 horsepower and finally 50 horsepower units. This was found necessary for unassisted takeoffs.

The landing gear started out with two large wheels just behind the center of gravity, with a pair of smaller wheels well behind, near the aft wing trailing edge. These were obviously intended to prevent the propeller from ground contact. Up front was a long skid to protect the foreplanes, and to serve as a brake on landing.

Subsequently, the aft smaller wheels were removed, thus permitting the aircraft to rotate for takeoff and landings. Had the front skid been a wheel, the first tricycle landing gear would have emerged.

It is worthy to note that this gear did permit the aircraft to be landed at higher angles of attack, and subsequently more slowly. The Wright brothers' longitudinal skids, designed for use in sand, encouraged more level landings. Actually the Wrights did not adopt wheels until their models B and R of 1910, both of which were "conventional" aft tail aircraft.

The long forward moment arm and the generous area of the front surfaces made the 14 bis longitudinally stable. However, the generous dihedral and interplane surfaces of the aft wings, made it readily susceptible to gusts and crosswinds. The 14 bis was a calm weather airplane. In light of today's hindsight however, the 14 bis did not add significantly to aviation history.

Santos Dumont continued his aviation career until 1910 and developed the popular Demoiselle conventional ultralight monoplane. After 1910, Santos' health declined and he blamed himself for the death and destruction wrought both by the airplane and the airship during World War I. The use of aircraft in the civil war in Brazil was the final blow. He committed suicide on July 23, 1932 at age 59 in Brazil.

ROE I BIPLANE—GREAT BRITAIN 1907

Mr. Alliot Verdon Roe became a famous designer and manufacturer of aircraft under the name of "AVRO." Especially fondly remembered is his AVRO "Avian" biplane trainer, closely resembling the De Havilland "Gypsy Moth" of pre World War II vintage.

In 1907 he developed the Roe I Biplane, a canard of which comparatively little is known. Its wing span was 30 feet, length 23 feet and it weighed 650 pounds. This machine was equipped with a pusher propeller on a long extension shaft, and was originally powered by a nine horsepower 4 cylinder J.A.P. motorcycle engine that was later replaced by a borrowed 24 horsepower Antoinette engine. He then changed the propeller and added small wings between the main wings at the outer ends, possibly for lateral control.

The landing gear was a four wheeled affair, with small wing tip wheels. The foreplane was a monoplane surface and both fore and aft wings were generously dihedralled. Since there was no fin or rudder aft of the main wings, Roe must have relied on the dihedral for directional stability.

Just what controls were used is shrouded in the mists of time. The machine made a short flight on June 8, 1908, and managed to rise two or three feet under its own power. For some time, this qualified as the first successful flight in England, but it was decided in 1928 that this short "hop" did not represent true flight.

Roe went on to found A.V. Roe and Company

Limited (AVRO). One of their most famous aircraft was the Avro Lancaster four engine bomber of World War II.

THE SILVER DART: CANADA'S FIRST AIRPLANE 1908

The Silver Dart was a joint Canadian-USA venture. It was developed by the Aerial Experiment Association. The prime-mover of this group was Dr. Alexander Graham Bell, inventor of the telephone, and his wife; Glenn H. Curtis, designer of motorcycle engines and later of flying boats, and Lieut. Thomas A. Selfridge who was killed in an airplane accident at Fort Myers while a passenger with Orville Wright. All were Americans Also involved were two young Canadians, Frederick Walker (Casey) Baldwin, and John McCurdy.

The Silver Dart, while based on earlier experiments, was largely developed by McCurdy in 1908. With Curtis, he developed an eight cylinder 50 horsepower water cooled engine. It turned a propeller on an extension shaft, via V-belt drive.

The airplane itself borrowed much from the Wright brothers, but had some unique features. The foreplane was a biplane structure which could be controlled in pitch. The main planes were also biplane, and a single controllable vertical surface at the rear served as a rudder. The main wings had dihedralled bottom surfaces, and anhedralled top surfaces. The pilot sat just forward of the main wing.

The unique features were triangular wing tips, hinged at their leading edge (ailerons), and a tricycle landing gear using motorcycle wheels—the nose wheel was steerable through a rudder bar, and was linked by cable to the rudder. The ailerons were also controlled by cable, but from the elevator control wheel, and were actuated by rotation of the wheel. The elevators responded to

Fig. 2-10. McCurdy's Silver Dart was the first airplane to fly in Canada. It was a product of the Aerial Experiment Association headed by Alexander Graham Bell.

Fig. 2-11. Horatio Barber's Valkyrie was not only a canard, but probably the first monoplane to fly in Great Britain. The 1910 aircraft was powered by a four cylinder, 35 hp Greene engine.

a push-pull rod from the same wheel. Most light airplanes of today are similarly controlled. The wings were covered with rubber faced silk, silver in color, hence the name "Silver Dart."

In February 1909, McCurdy flew this airplane for a half mile across the frozen surface of Baddeck Bay, Baddeck, Nova Scotia, Canada. It was Canada's first airplane flight. Two weeks later, McCurdy flew a distance of 30 miles in 24 minutes, circling to stay in the confines of the bay, for a landing on the ice.

The Silver Dart was 49 feet one inch in span, 30 feet long, had a wing area 420 square feet and, weighed loaded 860 pounds. Wing loading was 2.04 pounds per square foot of wing area.

During World War I, McCurdy operated an aircraft factory and a flying school to train young Canadians who volunteered for the British Royal Flying Corp and the Royal Naval Air Service.

After World War I, he became a world figure in aviation circles. During World War II, he became supervisor of purchasing and assistant director of aircraft production. In five years, Canada became the fifth largest allied air power and was "The Airdrome of Democracy." Some 131,553 aircrew were trained.

McCurdy was awarded the Order of the British Empire, and appointed Lieutenant Governor of Nova Scotia. The Canadian Government then dedicated a stone cairn bearing a bronze plaque commemorating "the first military demonstration of aircraft flight in Canada" at Petawawa, Ont. It named McCurdy and Baldwin and praised them for their genius, courage, and skill.

In 1959 McCurdy witnessed the 50th anniversary of his first flights; the flying of a replica of the Silver Dart by Wing Commander Paul Hartman DSC AFC. On October 25, 1960 the replica was put on view at the Aviation Museum at Uplands Airport, Ottawa, Ont. McCurdy passed away June 1961.

ASL VALKYRIE—GREAT BRITAIN 1910

Horatio Barber's Valkyrie single seat canard monoplane was an intriguing design with many features that differed from its contemporaries. It had a wire braced monoplane aft wing, with a reduced chord between the fore and aft framework in which the propeller operated. Short, wide ailerons were located at the wing tips. These hung vertically downward when the aircraft was at rest.

The forward surfaces were biplane. The upper, larger surfaces were fixed in flight but ground adjustable for longitudinal trimming. The lower, shorter surface was moveable in flight for pitch control.

The landing gear had a total of six wheels: four large ones, located in pairs under the longitudinal structure, plus two tail wheels of smaller diameter.

Twin split rudders were mounted behind the fore and aft structure. A central opening in each permitted clearance for the wing trailing edges as the rudders rotated on their vertical hinges.

The engine was a 35 horsepower Green four cylinder in-line. The wingspan was 34 feet, length 21 feet, and it weighed 520 pounds. Maximum speed was 35 to 50 mph.

Early canard designers had serious problems in correctly locating the center of gravity of their creations. Since engine and pilot weighed a substantial proportion of the aircraft's gross weight, their location, in a fore and aft sense, was critical to correct CG placement. This required that both pilot and engine be placed close to the main wing leading edge. If propellers were to be located aft of the wing's trailing edges, something obviously had to be done.

To accomplish this feat, the Wrights, Santos Dumont, Roe and McCurdy employed extension shafts, either chain and sprocket, V-belt or directly driven from the engine's output shaft. The Valkyrie's designer chose to mount his propeller directly on the motor shaft, in pusher configuration, rotating in the space provided by the shortered wing chord, as mentioned.

Judging from the main wheel location—which should be ahead of the CG for ground stability—the engine was just about on the center of gravity with the pilot seated just ahead; a very dangerous

Fig. 2-12. This French machine was built by Lefebvre in 1910. It featured coaxial, counter rotating props.

33

Fig. 2-13. Louis Bleriot, first to cross the English Channel in 1909, was very active in research and development. This machine, his Type 25, was an experiment with the canard concept. Was powered by a 50 hp rotary engine.

location in the event of a crash. Interestingly, it was typical of the ingenious Wright brothers to offset their engines to a position along side the pilot to avoid such possibly fatal injury.

Returning to the Valkyrie's ailerons, their vertical, at rest, position indicated that only the outer aileron, in a turn, was operated. The inner aileron then floated upward due to the higher angle of attack of the wing as a whole, necessary to offset the centrifugal force and in so doing reduced the inside wing lift. Adverse aileron yaw must have resulted.

The twin, short-moment-arm (to the C.G.)-rudders seem small, but it must be observed that there was very little lateral, destabilizing area ahead of the C.G. A later version, the Valkyrie "B," carried the rudders further back on extensions.

The Valkyrie was a slow but stable flyer. A further odd feature was that the ailerons were controlled by foot pedals while the twin rudders were stick-operated-by side to side movement, the opposite of today's convention. The forward, lower wing was controlled by fore and aft stick motion.

First flown in June 1910, the Valkyrie later made many successful flights.

LEFEBVRE-FRANCE 1910

Very little is known of this coaxial, counter rotating propeller machine.

Judging from the sketch this was a two place airplane of canard configuration. The engine and gear box were in an enclosure ahead of the pilot driving the propellers, which were set into the wing trailing edge. The landing gear was four wheeled, with an aft central skid for propeller protection.

Longitudinal control was achieved by angle of attack change of the foreplane outer portions. Ailerons, at the wing tips, provided lateral control. Directional control seemed to consist of a "rhino horn" fin atop the foreplane that could be rotated on a vertical axis.

While the wheels just ahead of the props added some aft lateral area, the impression remains that this canard must have been directionally unstable.

BLERIOT TYPE 25 - FRANCE

By 1911, Louis Blériot was prominent in French aviation circles. He is perhaps best known for his crossing of the English Channel from France, ending Great Britain's centuries old freedom from invasion except by the sea. He was an established producer of conventional aircraft world-wide, yet

34

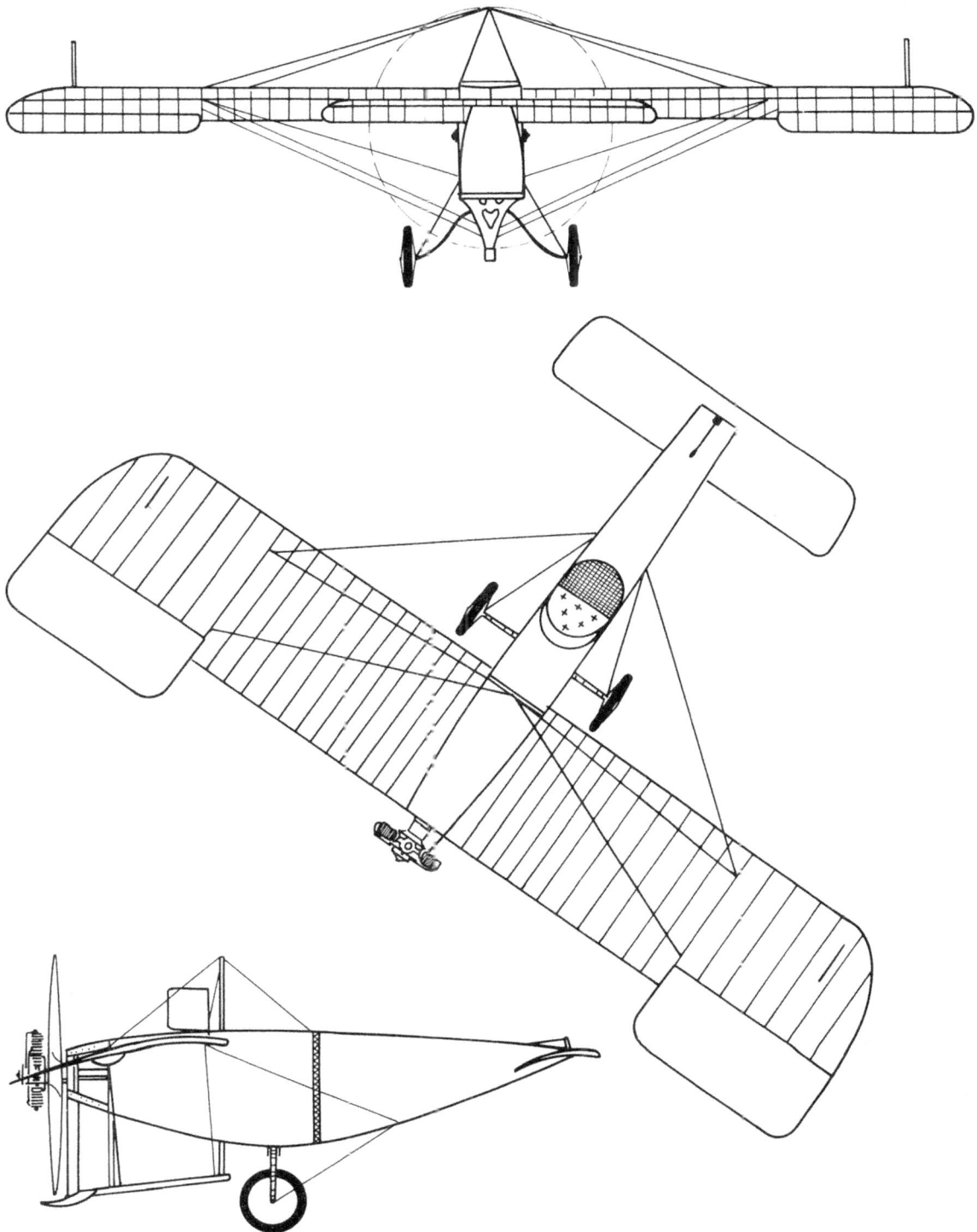

Fig. 2-14. Three view drawing of Bleriot's Type 25 canard.

he enjoyed experimental work. His type 25 was just that, and a reversion to the canard configuration. This aircraft was powered by a 50 horsepower Gnome rotary engine, turning a wooden propeller that was fastened to and in front of the engine. Both engine and prop rotated as one, mounted on a stationary shaft secured to the fuselage structure.

The rotary engine was noted for its gyroscopic effects, precession resisting sideways movement and causing strong nose up or nose down action, depending on the direction of the turn. In addition, the engine could not be throttled—the only way power could be reduced was by "blipping" the engine.

The aircraft spanned 30 feet and weighed 880 pounds. The wire spoked landing gear wheels were mounted on leaf springs reminiscent of automobile practice. A substantial tail skid provided ground stability and prop protection. The wide aft

struts must have provided some directional stability; the postage stamp size vertical rudders were not impressive. Blériot's 1913 version did have the much needed additional fin area.

Lateral control was provided by conventional ailerons, torque tube operated, that hung down vertically when the machine was at rest. Longitudinal control was achieved by the all-moving forward wing whose incidence was under pilot control. Landings, without pitching forward on the nose, must have called for "full-up" on the canard wing—particularly with a heavy pilot.

RAAB-KATZENSTEIN "RAKETE" — GERMANY 1929

The "Rakete" is another obscure canard aircraft. It was initially developed for rocket power. For initial flight test purposes however, it was equipped with a small 3 cylinder radial engine mounted aft of the main biplane wings.

Fig. 2-15. The German Raab-Katzenstein Rakete of 1929, was originally conceived for experiments with rocket power.

36

Fig. 2-16. The German Focke-Wulf Company developed this design, the Ente, in the late 1920's. It carried a pilot and three passengers, and proved to be a successful experiment in a stall-proof aircraft.

The landing gear was an unsteerable tricycle. The pilot rode in a streamlined enclosure from which projected a tubular structure that carried the foreplane. The latter was a double surfaced unit equipped with trailing edge elevators for pitch control. Roll and yaw controls were conventional.

Noteworthy, is the large amount of vertical surface area—in contrast to Blériot's type 25 canard. The central fin served to support the upper wing, while extending upward and aft. The outboard interplane struts also carried vertical surfaces.

Little is known of its weight, performance not its use as a rocket powered aircraft.

FOCKE-WULF "ENTE" (DUCK) 19a COMMERCIAL AIRPLANE - GERMANY 1930

In 1925, the Focke-Wulf Company persuaded the DVL (German Experimental Institute for Aeronautics) to construct this canard airplane. After extensive model tests, the Ente was built at the Gottingen Experimental Institute. Test flights proved excellent, but tragically, the plane crashed in September 1927 killing George Wulf, co-founder of the company. Since the accident did not relate to any fundamental design error, and after further wind tunnel tests, construction of a new Ente started, but it was destroyed in a fire in late 1929.

Finally, in May 1930, another new Ente was ready for flight trials which proved very satisfactory. The DVL fully accepted the airplane for carrying passengers.

The Ente was stall-proof. Even though the forward wing could be stalled, the aft wing could not. Therefore, lateral stability and control could not be lost. The foreplane was more heavily loaded than the main wing, and the center of gravity was

37

Fig. 2-17. Three view drawing of Focke-Wulf Ente.

located ahead of the main wing leading edge.

It was claimed that the Ente performed as well as the best commercial airplanes of the conventional rear tail configuration—afterall, it had no negative lift producing aft tail.

The Ente was a high wing cantilever monoplane. The wing had a thick profile, slight dihedral and "Zanonia" tips. It was claimed to have excellent lateral stability characteristics.

The Ente was demonstrated in several European countries before going to the DVL for aerodynamic research.

The foreplane was triangular in shape, with a straight leading edge, resulting in a forward sweep of the ¼-chord point. This wing planform characteristically starts to stall at the root trailing edge, with the stall progressing toward the wing tips as the angle of attack is increased. This would provide a gentle stall of the foreplane which is probably why it was selected.

The elevators, which provided pitch control, were actually slotted flaps suspended on brackets. The forward wing was ground-adjustable for trim, and it could also be tipped laterally about its center-line to provide a sideways force to offset off-center thrust in the event of an engine failure.

Directional control was provided by a huge fin and rudder at the fuselage rear, aided by small vertical surfaces beneath the wings.

Two 110 HP Siemens SH-14 air cooled 7 cylinder radial engines provided power. These were located beneath the aft wing, tractor fashion, with the propeller plane-of-rotation ahead of the main wing leading edge. Nacelles provided streamlining aft of the engines.

The landing gear was tricycle, to which the canard configuration lends itself very well. The main wheels were located beneath the engines, and the forward steerable wheel was partially submerged in the forward fuselage. At rest on the ground, the airplane sat "nose low" roughly 6 degrees. While this posture might have lengthened the takeoff run, it meant that the main wing was near its zero-lift angle, so that on landing the aircraft would cling to the runway as soon as the nose wheel made contact.

As was typical of the time, the pilot sat in an open cockpit, while the three passengers were accommodated in the cabin just behind the pilot.

The Ente was 32.8 feet in span, and 34.5 feet in length, with an empty weight of 2,590 pounds. It carried a load of 1,047.19 pounds—40% of its empty weight. Wing loading was 9.52 pounds per square foot and power loading 16.31 pounds per horsepower. Total wing area was 382.12 square feet.

Maximum speed was 88.2 mph, cruise 79.5 mph, and landing 51.6 mph. No high lift devices were employed. Controls were normal, with a separate hand wheel to tilt the forward wing.

GEEBEE "ASCENDER" - U.S.A. 1931

The Granville brothers of Springfield, Mass., along with chief designer Robert Hall, were famous for their line of GeeBee aircraft. All were fast, wire braced low wing monoplanes with well streamlined landing gear. All were powered by

Fig. 2-18. The Granville brothers of Tompson Trophy racing fame, produced this interesting single-seater in 1931 It proved to be quite stable, but suffered from inadequate aileron response due to its stock Aeronca main wing.

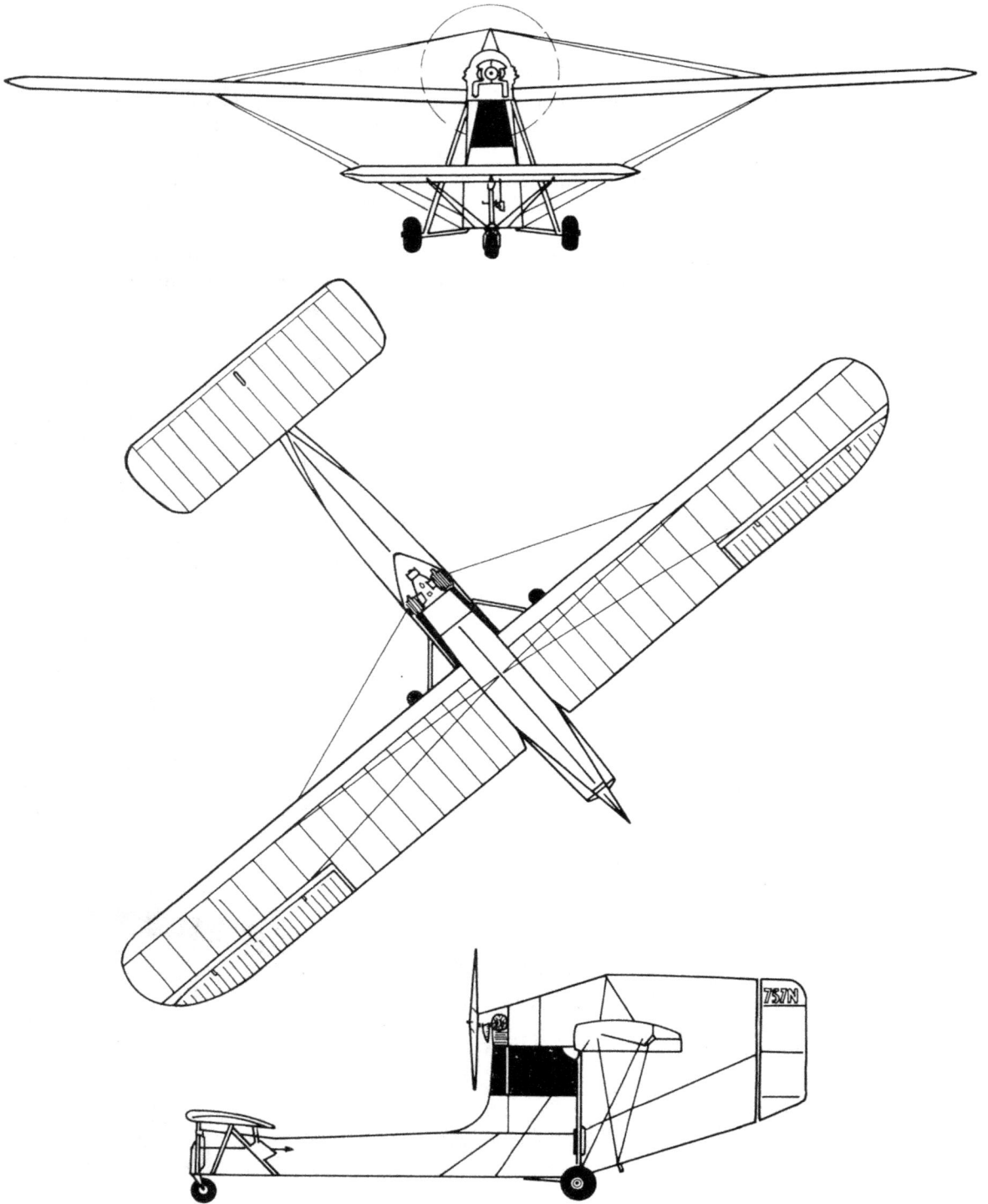

Fig. 2-19. Three-view drawing of Granville canard.

radial engines, except for the Model Sportster "D" which had an in-line, inverted, air cooled engine. Most famous were their "beer barrel"-shaped models Z and R1, both of which won Thomson trophies at the National Air Races in Cleveland.

Pilot Lowel B. Bayles was killed in model Z, which had been souped-up with a 750 wasp senior radial engine, in an attempt on the world's speed record at Detroit on December 5, 1931. GeeBee Model R6 QED participated well in the London—Melbourne (Australia) MacRobertson air race, but was the last GeeBee design.

The one "odd-ball" in the GeeBee line was their Ascender—a small single place canard made in 1931. Built from the remains of an Aeronca C2 in about one week, at a cost rumoured to be $500, the Ascender was powered by an Aeronca 2 cylinder engine of 28 horsepower. This plane was unique in its engine placement, just above and ahead of the wing. With the prop rotating about 1 foot ahead of the completely enclosed cockpit, this location provided near excellent visibility.

To clear the prop, the forward fuselage was shallow, and of triangular cross-section. The fuselage ended in a huge fin and rudder, which also served as anchor for the upper wire wing bracing. The aft fuselage lower surface was swept up sharp-ly to permit rotation on both takeoff and landing, without ground contact.

The foreplane was a fairly heavily cambered section that pivoted as a whole for pitch control.

The landing gear was tricycle, with the rear wheels located below the wing leading edge. The steerable nose wheel under the front of the fuselage, provided a long wheel base. Excellent landings were obtained.

The machine in its original form flew stably and was to fly, except for the Aeronca wing ailerons which gave a very sluggish response. Later on, an up-front rudder was installed on tubing running upward from the canard wing. This forward rudder was tested, both separately actuated and moved in concert with the rudder. It was subsequently removed.

On New Year's Day Mark Granville, a young and inexperienced pilot, banked the aircraft so steeply—and presumably at low altitude—that the limited aileron control did not permit recovery. The aircraft spiralled into the ground, inflicting a painful back injury to the pilot.

Since very little time and money were expended in its design, no formal drawings were produced. Unfortunately, the Granville brothers did not pursue their promising canard developments any further.

Fig. 2-20. Mignet's Pou du Ceil (Flying Flea) became quite popular after it was introduced in France in 1933. It created quite a surge in flying popularity, but unfortunately suffered from pitch instability due to the wings being too close to each other.

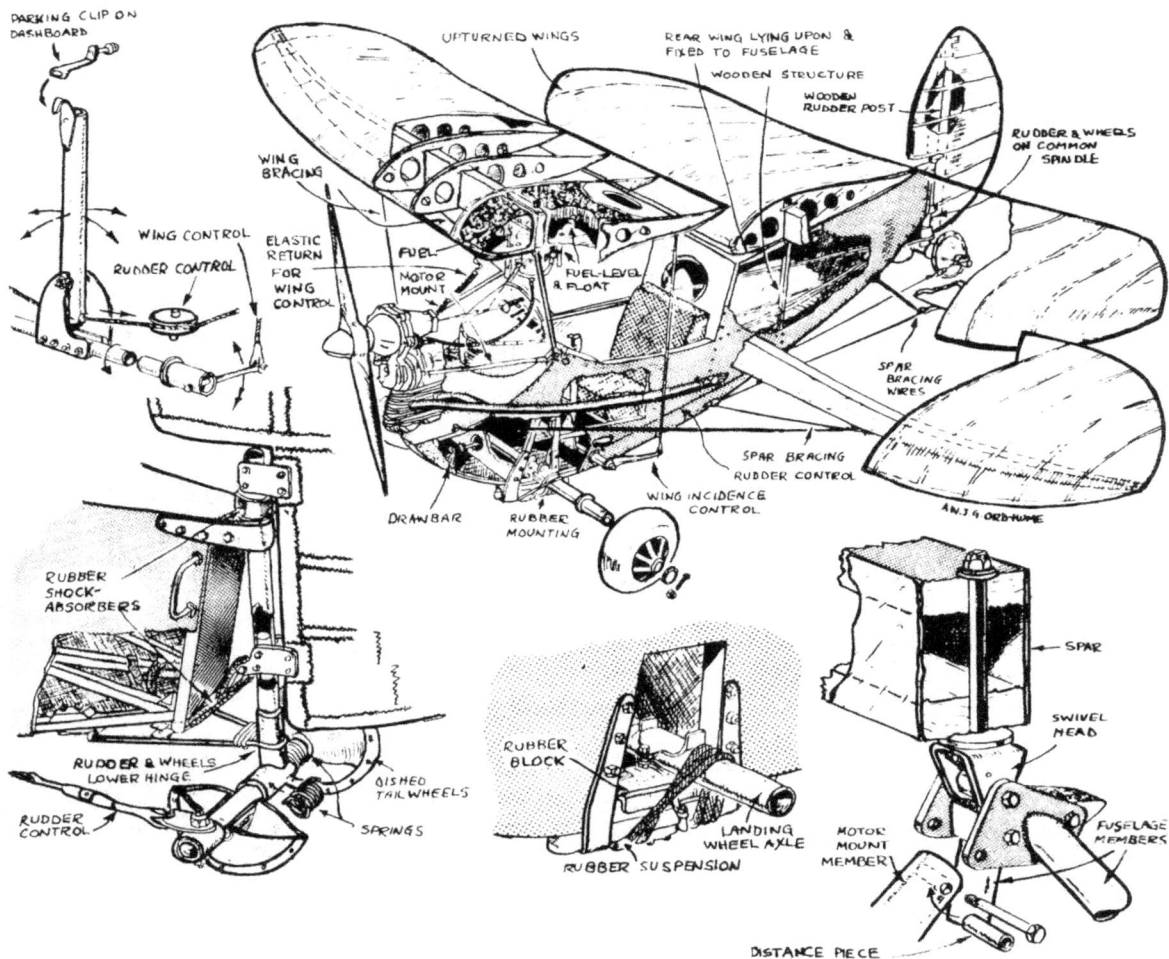

Fig. 2-21. Cutaway reveals details of Mignet's Pou du Ceil of 1935. Aircraft featured tandem wings. Entire front wing pivoted for pitch control.

MIGNET "POU-DU-CIEL" (FLYING FLEA) TANDEM - FRANCE 1933

The tandem wing has had a persistent popularity in France. Peyret and Albess in the "twenties;" Mignet, Mauboussin and Delanne in the "thirties;" Lignel in the "forties;" Clement in the "fifties;" Moynet and Desprets in the "sixties." Others too, developed versions of tandem wing aircraft.

Of all these, Mignet's "Pou du Ciel" or Flying Flea was the most popular. It was promoted as a small, low powered and inexpensive, homebuilt aircraft. It had a low horsepower air cooled motor mounted uncowled on a slanting firewall in the nose. The wings were close together longitudinally—the front wing trailing edge just overlapping the rear leading edge, with a narrow vertical gap between them.

Pitch control was by changing angle of attack of the forward wing. Since the pivot point was well forward, this meant a load on the control column at all times in flight, increasing with speed. The trailing edge of the forward wing was connected by cable to the control stick. A bungee arrangement kept the wing trailing edge from bouncing off the pilot's head while taxiing.

Sideways movement of the stick actuated the huge all-rudder vertical surface, causing a yaw, so

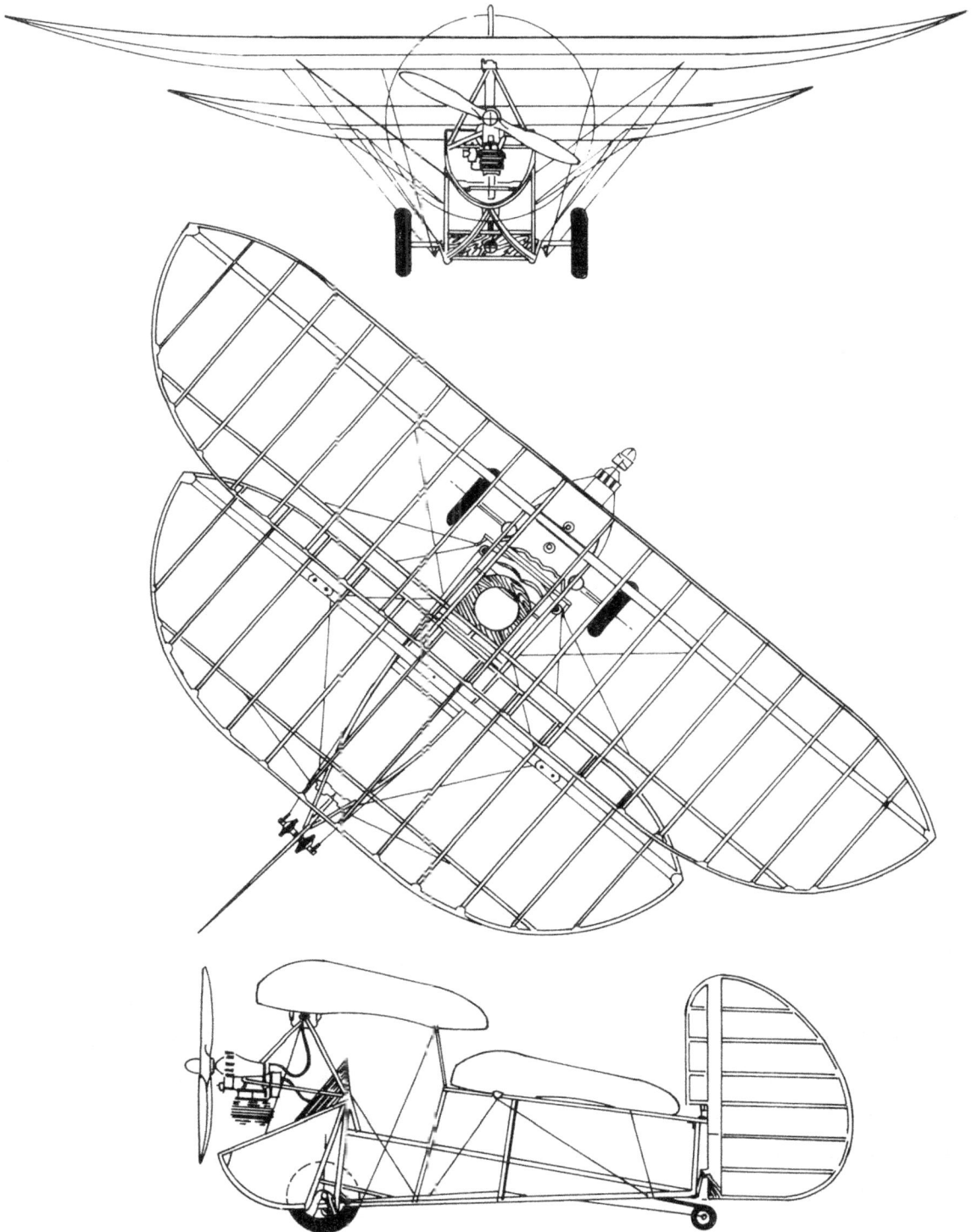

Fig. 2-22. Three view drawing of the Pou du Ceil, or Flying Flea.

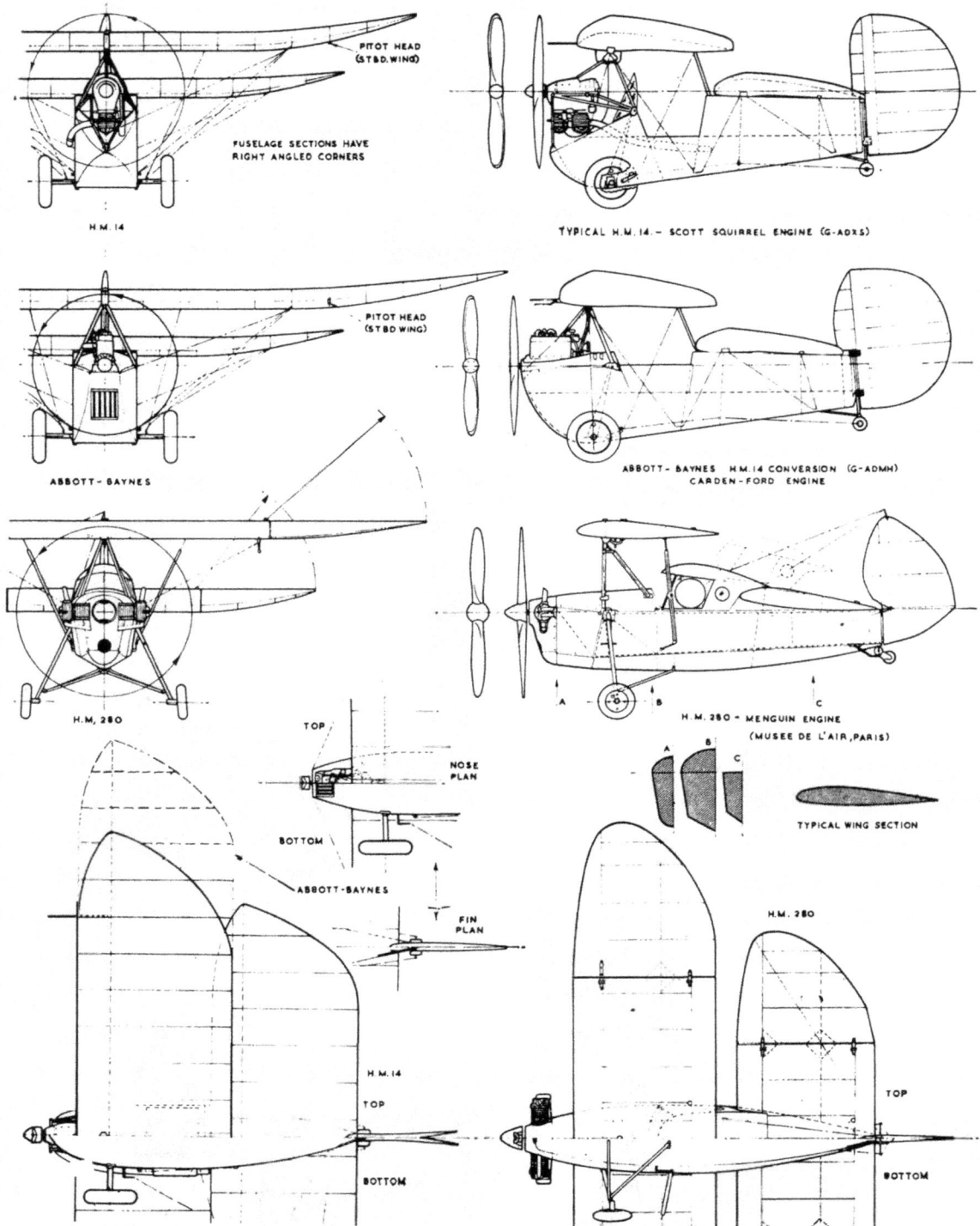

PITOT HEAD
(ST'B'D. WING)

FUSELAGE SECTIONS HAVE
RIGHT ANGLED CORNERS

H.M. 14

TYPICAL H.M. 14. – SCOTT SQUIRREL ENGINE (G-ADXS)

PITOT HEAD
(ST'B'D WING)

ABBOTT–BAYNES

ABBOTT– BAYNES H.M.14 CONVERSION (G-ADMH)
CARDEN-FORD ENGINE

H.M. 280

H.M. 280 – MENGUIN ENGINE
(MUSEE DE L'AIR, PARIS)

TOP

NOSE
PLAN

BOTTOM

ABBOTT-BAYNES

A   B   C

TYPICAL WING SECTION

FIN
PLAN

H.M. 280

H.M. 14

TOP

BOTTOM

TOP

BOTTOM

Fig. 2-23. Drawings trace the evolution of the Flying Flea tandem.

44

Fig. 2-24. Cutaway drawing reveals the wooden structure of the Mignet HM-36, a prototype built in 1942 in the United States.

that the faster side produced more lift and consequently banked the airplane. No ailerons or rudder pedals existed. Two control functions in a three dimensional environment! Due to the very limited flight envelope, pilots who exceeded this limitation, either in slow flight or at high speed, ended up in dives from which they could not recover. Many were killed.

It was found that, in low speed, pulling back on the stick steepened the glide; the rate of descent increased until the airflow over the front wing separated reducing its lift, but the aft wing, at lower angle of attack, operating in the foreplane downwash, continued to lift forcibly. A dive resulted. Pulling back the stick to raise the nose only aggravated the situation.

As the speed increased, on the other hand, the front plane angle of attack must be gradually reduced to maintain level flight. This resulted in the aft wing supporting more and more of the aircraft's weight and a rearward shift in the center of lift occurred. The plane then proceeded to nose over. As speed increased so did the up load which had to be opposed by the pilot. Throttling back was the only recourse, if enough altitude was available.

The original lower powered "Pou's" had a limited range of useable speeds; climb, cruise and glide being only a few miles per hour apart. On the HM-14, however, the dive became so violent that back stick and closed throttle were of no avail, and down it went.

45

Wind-tunnel tests indicated a need for more angle of attack from a maximum of 9 degrees to about 14 degrees. Other changes included larger diameter leading edge radii, reduced wing overlap, replacement of the pull cable by a strut, and on a few models, addition of an aft-wing trailing edge elevator to provide more pitch control.

Fundamentally the same airplane is still flying today, however in more sophisticated versions, seating two and three people. Even so, the same limitation of pitch control and lack of ailerons, causes serious problems in cross wind takeoffs and landings.

BELTRAMO "COLIBRI" - ITALY 1937

Italy now enters the canard development field. Quinto Beltramo designed a very attractive little airplane, with many unique features. It was powered by an 18 horsepower single cylinder, two-stroke engine. With the propeller shaft mounted high at the rear, and a deep aft fuselage, intended to provide directional stability, the fuselage partially contained the aft wheel of a tandem landing gear.

The steerable forward wheel was fully exposed. The aft wing was of high aspect ratio and tapered in planform with a straight trailing edge, but had no ailerons and no rudder.

All flight controls were concentrated in the high mounted forward wing, which could be controlled, incidence-wise, from -2.5 to +12.5 degrees. It was also articulated laterally so that it could be see-sawed through 30 degrees on either side.

Tipping the foreplane sideways, plus a small incidence increase, produced a side vector intended to turn the aircraft's nose in the direction of the tip. Like the Pou du Ciel, the higher speed of the outside wing induced the bank—another example of two control functions in a three control environment.

Just what happened in test flying the Colibri is unrecorded; but judging from the absence of any further mention in the Aviation Press, of the machine or its designer, the results could not have been outstanding.

STEFANUTTI SS2 and SS3 - ITALY 1937

Sergio Stefanutti developed his SS2 in 1935. Powered by a 16 horsepower two cylinder Keller engine, the comparatively orthodox design was quite successful. It was a very well streamlined aircraft with tapered aft wings, ailerons, and wing mounted rudders.

The wing was set low in the fuselage, and the well faired wide tread main wheels were located just behind the wing leading edge. The engine

Fig. 2-25. A 1937 Italian experimental canard by Quinto Beltramo. It was powered by an 18 hp, single cylinder pusher. All control functions were handled by the canard: incidence changes controlled pitch, while sideways tilts controlled yaw/roll.

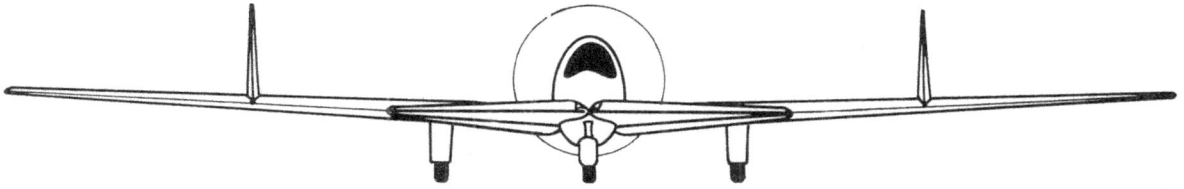

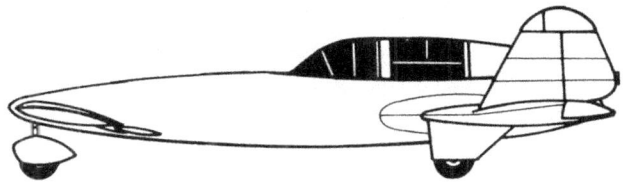

Fig. 2-26. Sergio Stefanutti's SS2 first flew in 1935, powered by a 16 hp engine. Three view drawing depicts the 37 hp two-seat SS3, which could be considered modern by anyone's standards.

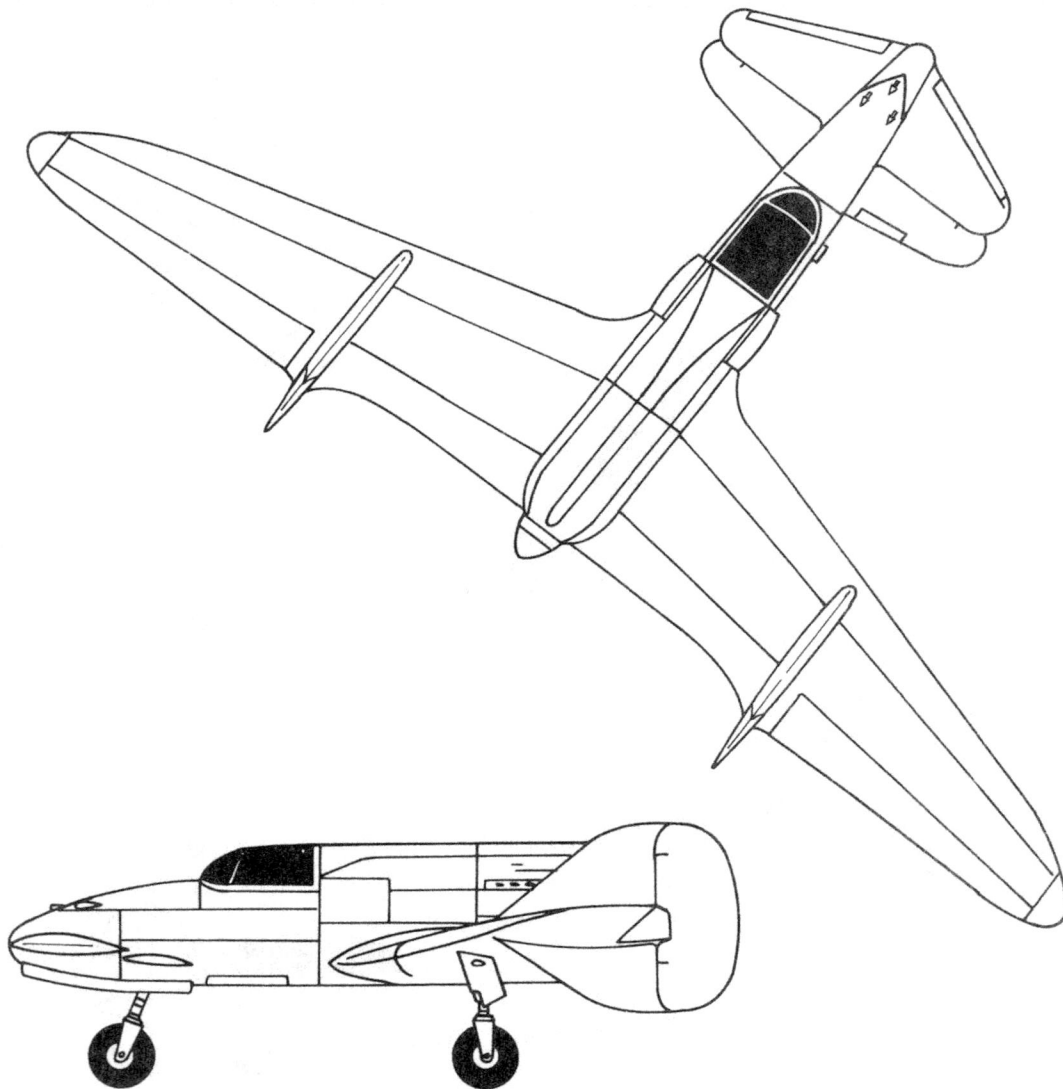

Fig. 2-27. Three view drawing of the Ambrosini SS4 fighter, designed by Sergio Stefanutti. Unfortunately, the right aileron fluttered and tore off, sending the aircraft into the ground. This ended Italy's involvement in the canard concept.

thrust line was high to provide ground clearance during the flair. The steerable nose wheel was enclosed in a streamlined pant and the pilot sat ahead of the wing with an excellent view.

The forward wing, of unusual ovoid shape in plan view, was set at a high angle of incidence. The elevator was of the external airfoil type located below and just ahead of the wing trailing edge, providing a slot effect.

This aircraft was the most sophisticated canard design to appear, and would be considered "modern" by today's standards. The SS2 was so successful that a two place version was developed, powered by a 37 horsepower CNA engine, which was exhibited at the Italian "Salone Internat onale Dell Aeronautica" in 1937. Stefanutti then went on to design the SS4.

SAI AMBROSINI SS4 - ITALY 1939

The SS4 was a piston engined canard fighter similar in configuration to the SS3. The wings were tapered and swept back, with straight trailing edges, recessed to provide prop clearance.

Large fins and rudders were wing-mounted inboard of the ailerons, and projected well aft of the wing trailing edge. The canard employed a fixed forward wing with an external airfoil elevator and slot arrangement, supported and articulated on brackets for pitch control

The pilot had an excellent field of view, and the aircraft was provided with machine guns above the canard, firing forward. The landing gear was tricycle and fully retractable.

The SS4 flew on March 7, 1939 for a five minute flight, but the following day's flight ended in disaster. The wing developed flutter, the right aileron tore off, and the aircraft dove into the ground.

This ended canard development in Italy.

DE HAVILLAND AERONAUTICAL SCHOOL TK-5 GREAT BRITAIN 1939

The students of this technical school began the design of a canard in 1938, obviously influenced by Stefanutti's SS3 which it closely resembled.

Powered by a 14S Gypsy Major engine, an inline, inverted, air cooled unit used in many English aircraft of the day, the TK-5 had a swept wing of beautiful proportions. Twin fins and rudders close to the wing tips, extending below the wing, along with the deep aft fuselage and main wheel fairings, contributed to directional stability.

The forward plane was quite small and was equipped with conventional elevators. It had a small but noticeable dihedral paralleling that of the wing.

The TK-5 never flew. As it accelerated for takeoff it clung firmly to the ground, possibly due to low elevator power; aft wheels positioned too far behind the center of gravity and the high thrust line, all of which prevented rotation.

The plane was completed in late 1939, and together with its drawings was destroyed before it was painted—undoubtedly by enemy action dur-

Fig. 2-28. Students of the De Havilland Aeronautica School designed this beautifully proportioned canard, the TK-5, in 1939. Unfortunately, it never flew, and fell victim to the outbreak of WWII.

Fig. 2-29. Three view drawing of De Haviland TK-5.

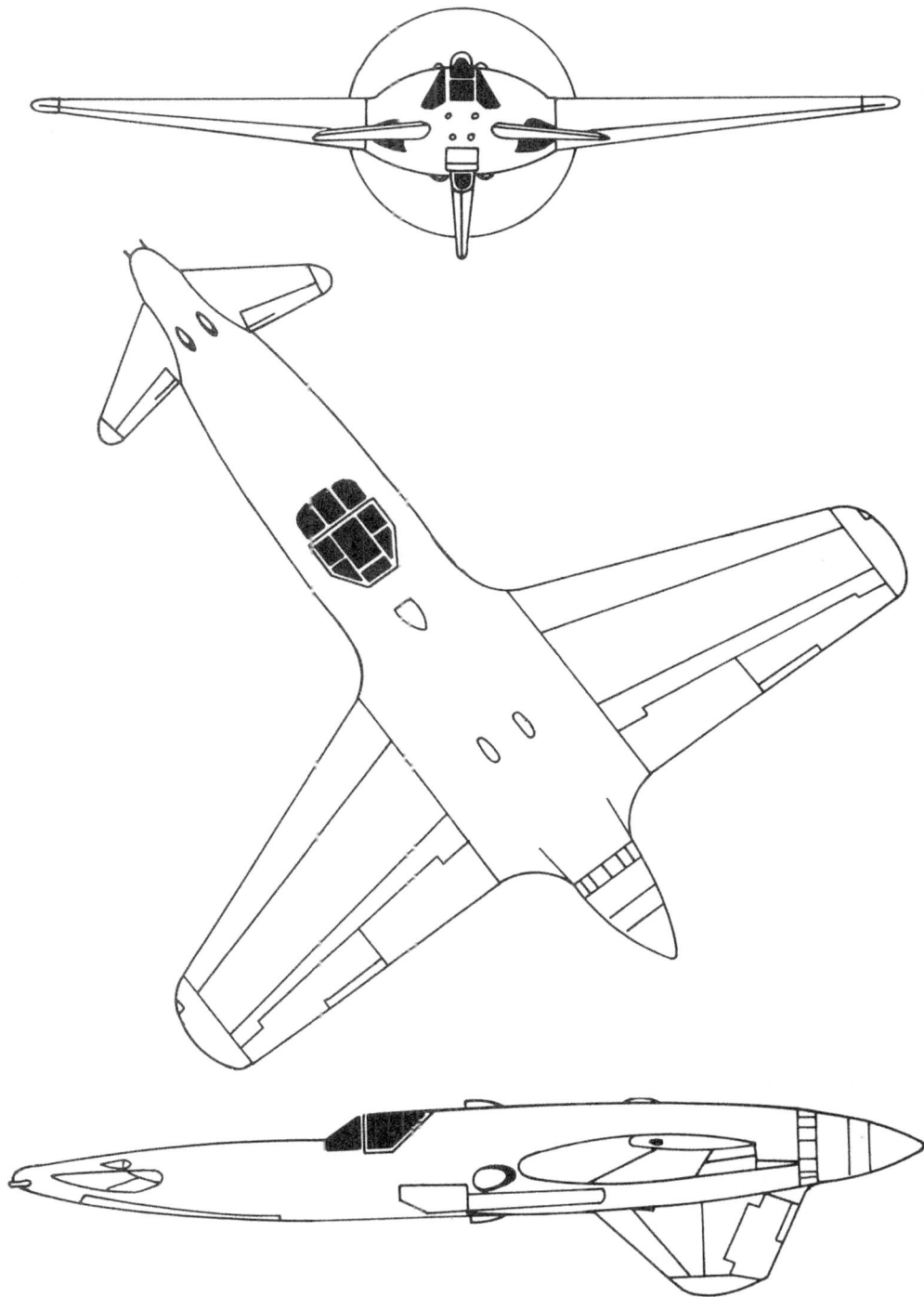

Fig. 2-30. Three view of the German Henschel P.75 heavy fighter. Two engines, mounted at the CG, drove contra-pusher props. Aircraft featured heavy firepower in the forward fuselage.

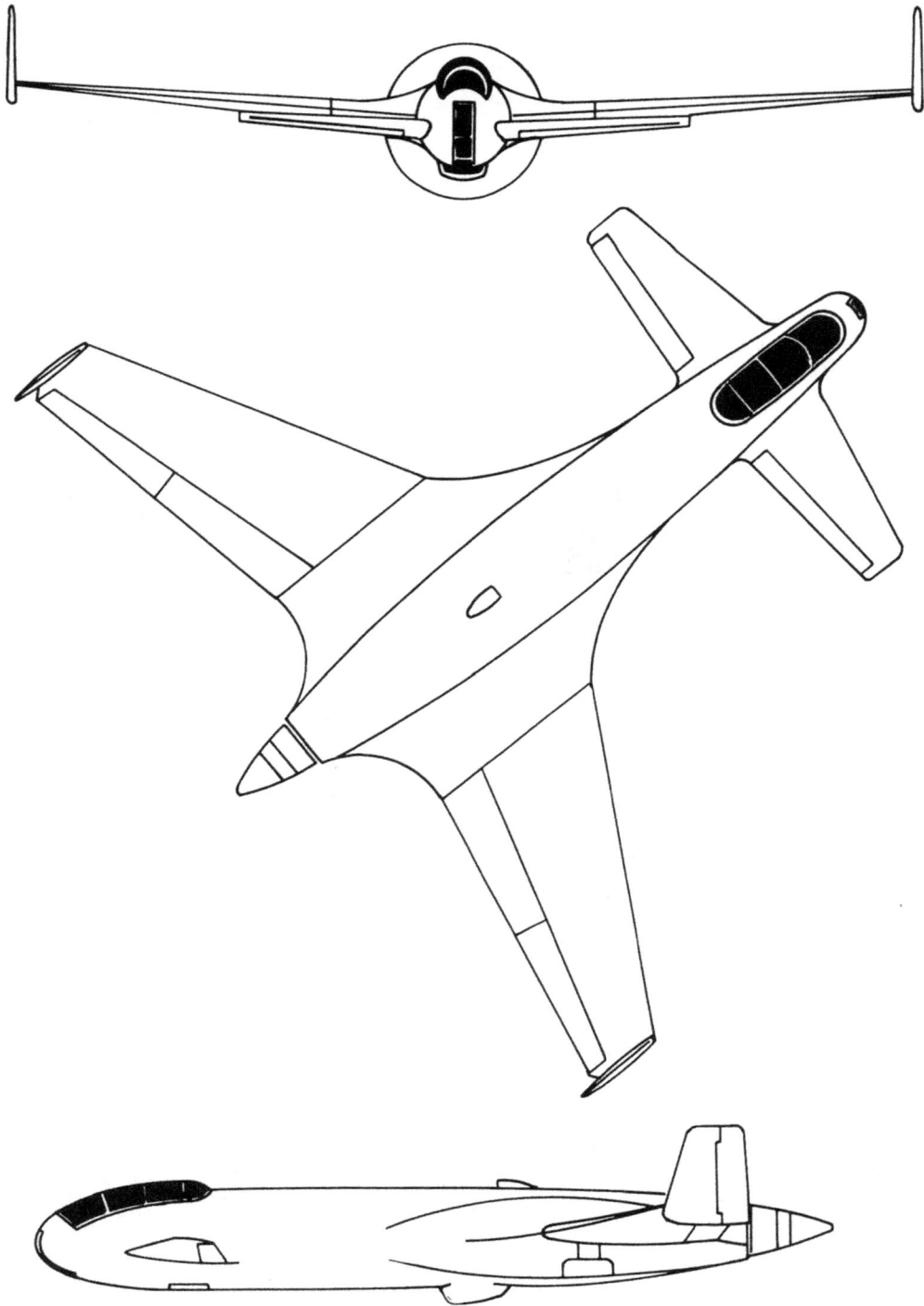

Fig. 2-31. Three view drawing of the Henschel P.87 high speed bomber of 1941.

ing World War II which broke out shortly thereafter.

HENSCHEL FLUGZEUGWERKE P.75 AND P.87 - GERMANY 1941

During World War II, the Germans investigated the canard, along with every other conceivable aerodynamic configuration. Some, like their Messerschmidt ME 163 tailless, rocket powered "one-shot," with four cannons and a duration of eight minutes made it into service. Neither the P.75 fighter or the P.87 high speed bomber got past the wind tunnel testing stage, however.

The P.75 had an ellipsoidal fuselage cross section blending into an upswept wing of relatively deep cross-section. Coupled DB 601 engines were installed at the center of gravity, driving six bladed counter rotating propellers through extension shafting. The retractable tricycle landing gear was quite long to assure adequate propeller ground clearance during landing and takeoff. A huge vertical fin and rudder provided directional stability and control. The forward wing was swept back and provided with elevators. The aft plane carried ailerons and flaps. This configuration had very concentrated fire-power in the forward fuselage; cockpit visibility was excellent and as the three view drawings show, it looked "lethal."

Henschel also developed the P.87 along canard lines, very similar to the P.75, but with a substantially larger foreplane, unswept but with a swept aft wing and wing mounted vertical surfaces. Unfortunately, the German War Ministry rejected both these interesting designs.

VICKERS TYPE "C" BOMBER - GREAT BRITAIN 1942

In late 1942, Vickers investigated canards or "Experimentals with Foreplane," according to Air Ministry specification B3/42.

This was a six British Centaurus engined monster with a wing span of 310 feet. It's total wing area was 2,900 square feet, its aspect ratio 15, overall length 96 feet, and an all up weight of 168,000 pounds. Estimated cruising speed exceeded 300 mph.

Vickers quoted the following advantages:

- Greater takeoff weight (5 to 10%) because the lifting foreplane eliminated the download of the conventional rear horizontal tail surfaces.
- Lower landing speed due to flaps on the foreplane.
- Due to long wheel base and wide tread, the tricycle landing gear would be excellent.
- Better longitudinal stability, since the foreplane was ahead of both wing downwash and propeller slipstream.
- Unimpeded rear field of fire for defense.
- A high aspect ratio aft wing, with narrow chord, resulted in reduced nose down movements from deployed trailing edge flaps. Longitudinal balance could be achieved by interconnecting high lift devices on the foreplane, so that both sets of devices operated simultaneously to obtain zero trim change.
- Suited for conversion to commercial use.

The British Air Ministry refused this design; their policy was against such large aircraft—production would have taken over three years. It was envisaged that World War II would be over in this time period and hence this canard bomber would no longer be required. The plane was never built.

Vickers, having decided that the canard configuration was more efficient aerodynamically, went on to design civil aircraft based on the type "C" configuration. Considering the heavy demands of the bombing offensive and the conventional bombers required, development of civil aircraft of unconventional design was discouraged.

This is the first canard in this review to incorporate high lift devices for landing speed reduction.

CURTIS-WRIGHT XP-55 "ASCENDER" - U.S.A. 1943

Curtis Wright conceived the XP-55 in response to the U.S. Navy Air Corps Circular Proposal R-40C, intended to promote radical new departures in a fighter design.

The XP-55 was a tail first, laminar flow wing sectioned, swept wing, all metal airplane, with retractable tricycle landing gear. It was powered by an Allison VI710 engine and pusher propeller. Small wing-mounted fins and rudders, plus deep

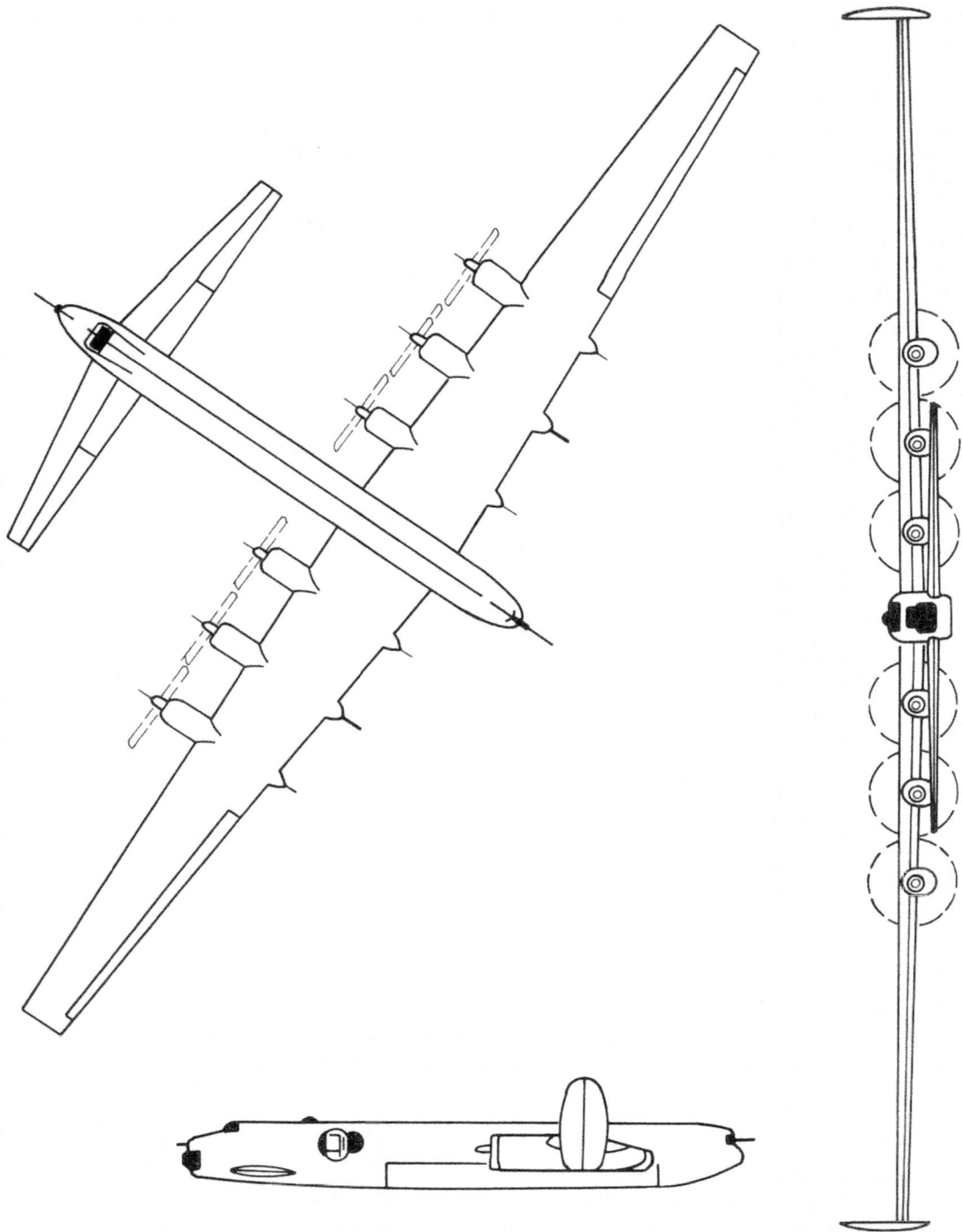

Fig. 2-32. England's Vickers Aircraft Company investigated the canard concept with their Type "C" bomber, according to Air Ministry Specification B3/42. This 1942 design was the first to incorporate high lift devices. Development was halted because it was felt WWII would be over before the aircraft could be operational.

Fig. 2-33. In response to US Navy Air Corps Proposal R-40C, Curtiss-Wright designed the XP-55 Ascender. Unfortunately, lack of complete research lead to its abandonment.

aft fuselage vertical projections, provided directional stability and control. In the event of an emergency requiring pilot bail-out, the aircraft was provided with an explosive device to blow the propeller off so that the pilot would not be literally, sliced as he ejected.

This tail first entry was initially turned down after wind tunnel testing of models because of poor stability and bad stall characteristics.

The author's analysis of the design, based on NACA documentation, indicates that the center of gravity was not somewhere between fore and aft wings; but was inside the limits of the aft wing mean aerodynamic chord (MAC), located roughly halfway between the 25% chord position and leading edge; just about where it would be on a conventional aft tail design.

Instead of a downward loaded horizontal tail plane, the XP-55 employed an upward loaded,

forward, free floating, tab operated, mass balanced stabilizer of symetrical cross section—certainly not a true canard wing.

Curtis Wright persisted and built a flying mock-up Model 24B, powered by a 275 horsepower Menasco engine, that was extensively test flown. The aircraft, despite problems, was considered worthy development. Three XP-55 prototypes were ordered by the US Army Air Force.

Flight testing started on the first prototype in mid-1943, and an excessively long takeoff run dictated a 15% increase in the front stabilizer area. Also, the aileron up-trim had to be interconnected with the flaps, so that up aileron coincided with the split flap deployment.

During stall tests with flaps deployed, the wing root section stalled. The plane pitched forward 180 degrees ending up on his back in a vertical descent—as predicted by the original army wind

55

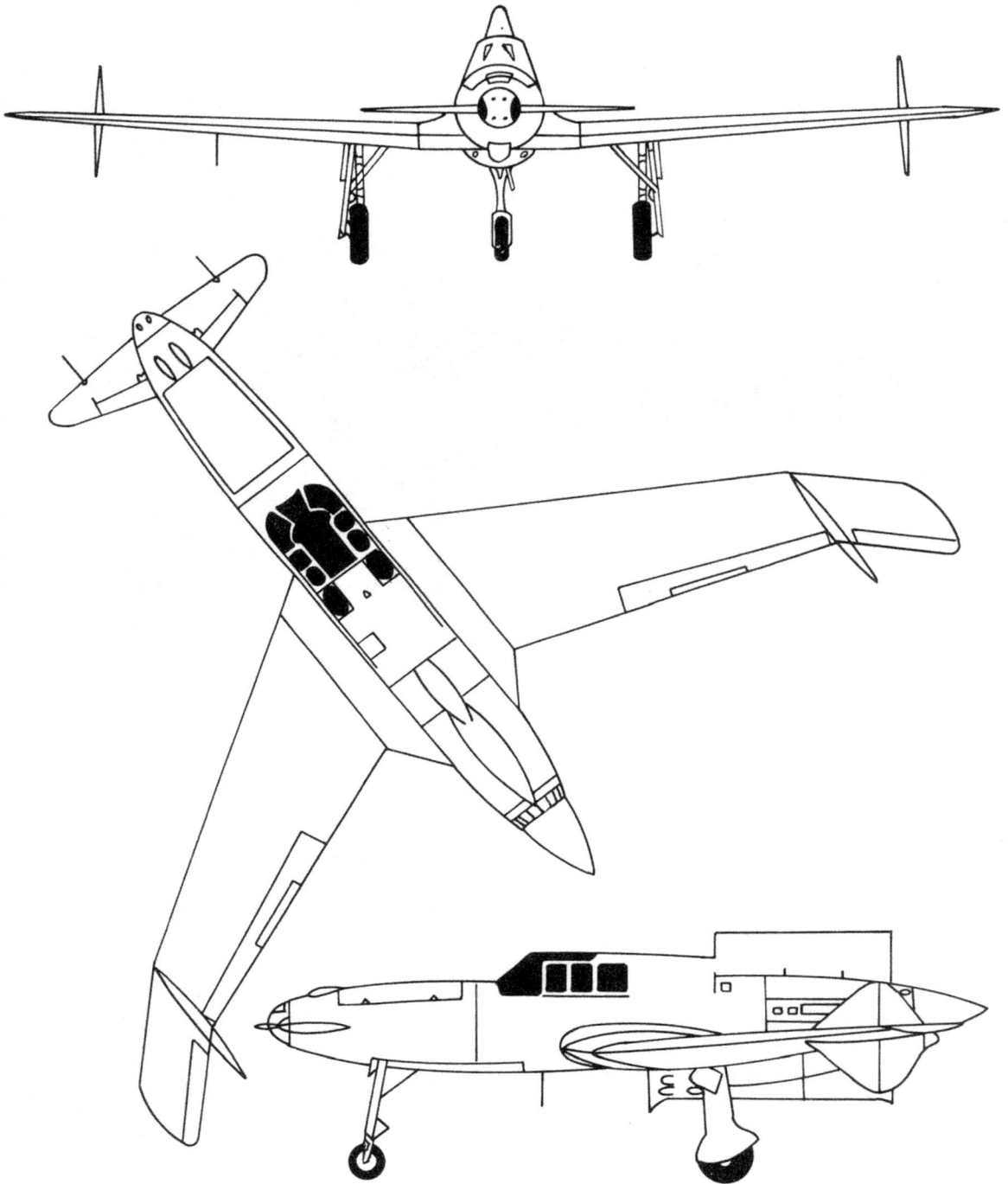

Fig. 2-34. Three-view drawing of Curtiss-Wright XP-55 Ascender.

tunnel tests. The explosive device for prop removal failed to operate; fortunately the engine, now inverted, quit and the pilot hailed out after falling 16,000 feet, trying to break the stall but to no avail. The plane continued straight down and was destroyed by the impact. The pilot survived.

Many things were tried to improve the stall. The wing tips were extended two feet, elevator travel limits were increased, and an inverted fuel system was installed to permit the pilot to "power" himself back into upright flight.

Flight testing with another prototype continued for much of the remainder of World War II. However, it was concluded that this "canard" aircraft, with its concentrated forward fire power and superb cockpit visibility did not offer sufficient advantages over the conventional to compensate for its bad stall and very poor stability.

The remaining XP-55 is on display at the National Air and Space Museum, Silver Hill, Maryland.

UTKA (DUCK) - RUSSIA 1944

Late in World War II, students at the Zhukovskii Academy supervised by Prof. G.A. Tokaev, developed the design of this small canard aircraft as a pure research effort.

Built of wood and fabric at the Mikoyan-Gurevitch bureau's experimental factory, the UTKA was powered by a 150 horsepower M-II radial engine. It could accommodate a pilot and two passengers.

The constant chord aft wing was swept back 20 degrees, mounted at the fuselage top and strut braced. Vertical surfaces at 55% span extended well behind the wing trailing edge. Leading edge slats ahead of the ailerons improved lateral control at high angles of attack.

A substantial fixed foreplane with tabbed trailing edge elevators, provided both lift and longitudinal pitch control. The landing gear was a fixed tricycle, with provision for both wheels and skis. The nosewheel was steerable. Removing the aft wing and its bracing, the aircraft could be converted to an air-sled on skis. The cabin provided excellent forward visibility.

It was test flown by Alexander Zhukov in 1945, but poor stability and other problems caused its abandonment after limited tests.

Fig. 2-35. In 1944, students of the Zhukovski Academy designed this small canard as a pure research project. Built of wood and fabric, the Russian Utka was powered by a 150 hp radial engine. Poor stability led to the abandonment of this interesting three-place aircraft.

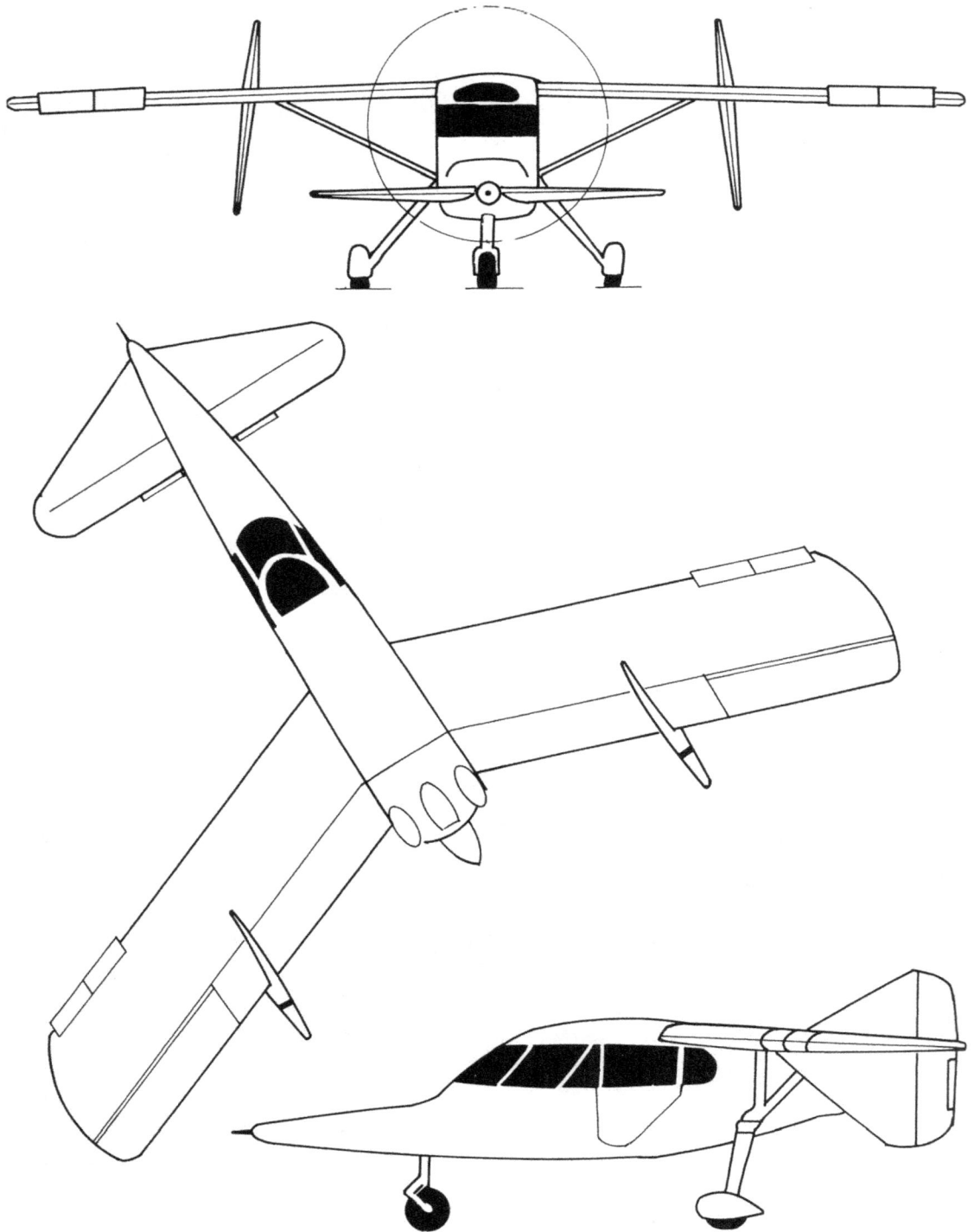

Fig. 2-36. Three view drawing of the Russian Utka.

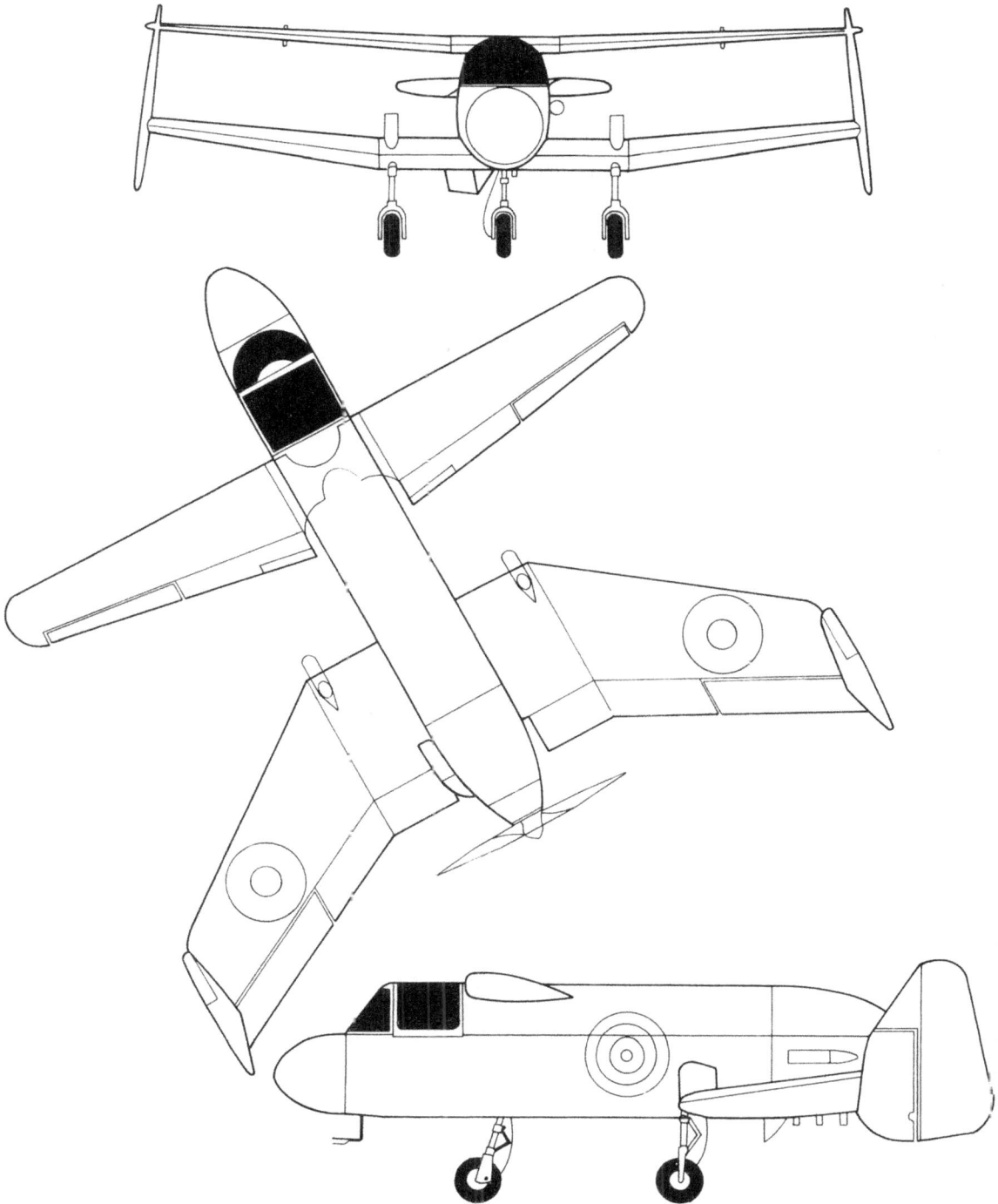

Fig. 2-37. The English firm of Miles, first flew their Libellula M.35 tandem in 1942. Both wings of this successful canard, incorporated flaps. Libellula is the entomological name for dragonflys.

Fig. 2-38. The Miles M.39B was built as a private venture, and sold to the British government. It was intended as a ⅝-scale research model of a projected bomber. The twin-engine aircraft weighed 2,600 pounds, and had a top speed of 164 mph. Longitudinal stability and control were satisfactory, but under wartime pressures, the Ministry of Aircraft Production discontinued further development.

MILES NO. 39B "LIBELLULA" -
GREAT BRITAIN 1945

Called a "tandem biplane," this airplane was designed and built as a private venture by Miles Aircraft Ltd., and was purchased by the British Air Ministry. Flight tests were performed at the Royal Aircraft Establishment, Farnborough, England, during 1944 and again in 1945.

The M39B was a single seat canard of wooden construction, equipped with retractable tricycle landing gear. Power was by two Gypsy Major MK16 140 horsepower in-line, inverted, aircooled engines driving fixed pitch propellers of 6.5 foot diameter. It was intended as a ⅝ scale model of a bomber; as a result, fortunately, of the flight tests and subsequent reports (Ref. 19) there exists extensive documentation on the airplane's design and flight characteristics, which is synopsized herein.

The plane weighed 2,600 pounds during the trials, had a span of 37.5 feet, and a length of 22 feet. The rear wing was swept back and fitted with plain flaps inboard and ailerons outboard. The front wing had elevators inboard and plain flaps outboard. Flaps on either wing could be lowered independently for test purposes. Lowering the front wing flaps caused the elevators to droop roughly one third of the flap angle of deployment.

Directional stability and control were provided by end plate fins and rudders, toed in one degree, mounted on the aft wing tips. A large central fin on the fuselage was found necessary for directional stability. Noteworthy, is that the center of gravity was located roughly 40 percent of the distance between the quarter chord points of the mean aerodynamic chords of the two wings, forward of the quarter chord point of the rear wing. In normal flight the front wing was lifting about 40 percent of the aircraft's weight.

Maximum level speed at sea level was 164 mph at a gross weight of 2,770 pounds. A minimum speed of 59 mph was experienced.

Handling characteristics were normal for an aircraft of this power and wing loading and did not differ from those of a conventional aircraft. Aileron control was sluggish at low speeds.

The amount of lift available on the front wing did not permit more than ten degrees of flap on the rear wing for landing. Longitudinal stability and control were satisfactory. At cruising speed and forward CG, a small amount of front flap was found necessary to achieve longitudinal trim. A slight increase in incidence of the front wing was recommended to give zero flap elevator angle in cruise.

Directional stability and control was satisfactory with the added central fin. Rudder control was satisfactory at all speeds, and good turns were possible on rudder only, with no difference between port and starboard application.

Ailerons, as mentioned, were sluggish and ineffective below 85 mph, particularly if the rear wing flaps were lowered. They were effective at high speed.

Good turns were possible on ailerons only with little slip going in or coming out. Displacement of ailerons produced a small amount of yaw (presumably favorable—due to differential ailerons). Lateral stability in a turn was neutral—the aircraft continuing to turn with ailerons neutral.

Straight sideslips of 15° at 110 mph, with about 40° of bank, were achieved. Three quarter aileron was required and the rudder was the limiting control.

Single engine flight was possible, but unpleasant due to low speed, bad aileron control for level flight and large rudder load required. Altitude could be maintained on one engine.

Ground handling, takeoffs and landings were straightforward.

Approaches indicated a need for higher lift on the foreplane, flaps down, to permit a reasonable 40° rear wing flap angle for lower flying speeds.

Despite the forward wing 1° greater incidence than that of the rear wing, the much greater wing loading of the front plane assured that it would stall first. The impact of the downwash off the front wing would have also lowered the effective angle of attack of that area of the main wing behind the foreplane, inducing more stall resistance at the main wing center section. Propeller slipstream would have also played a part.

Stall trials were performed at 5,000 to 6,000 feet altitude, after the aircraft was fitted with anti-spin parachutes in the aft wing root, although no mention was made in Ref. 19 to spin testing. CG locations in these trials were at the center of the normal CG range and later at a point two inches aft.

The first set of tests were made with no aft wing flap deployment, but with the front wing flap at angles from 0° to 44.5° and with "backstick" (up elevator). Bear in mind that front flap and elevator both move when the former is deployed—but the elevators only move one third of the flap angle.

With both front and rear flaps at 0°, and full up elevator (stick back, elevator down) the aircraft was not stalled and descended steadily at 81 mph, engines idling.

With increasing front flap angle of deployment and heavy backstick application, stalls occured at speeds from 69 to 63 mph. With increased front flap angle the stall became more pronounced and if backstick was still held, a pitching oscillation was induced that became progressively more severe. The stall and pitch oscillations could easily be stopped by firm forward stick or if the front flap was raised.

A second series of tests were made with full forward flap deployment, and progressively more aft flap, up to the maximum of 21° and progressively more backstick. Starting at 2/3 backstick while at lower aft wing flap settings of 4°, 9° and 13°, the nose dropped without any stall warning, but the pitch oscillation was easily avoided.

At 17° and 21° aft flap angles, the aircraft remained unstalled and descended steadily at 63 mph and 71 mph respectively.

The controls became progressively less effective, particularly the elevator, as the rear flap setting was increased. Obviously, there was just no remaining elevator power to level, and even less to flair the aircraft for landing. For this reason, aft

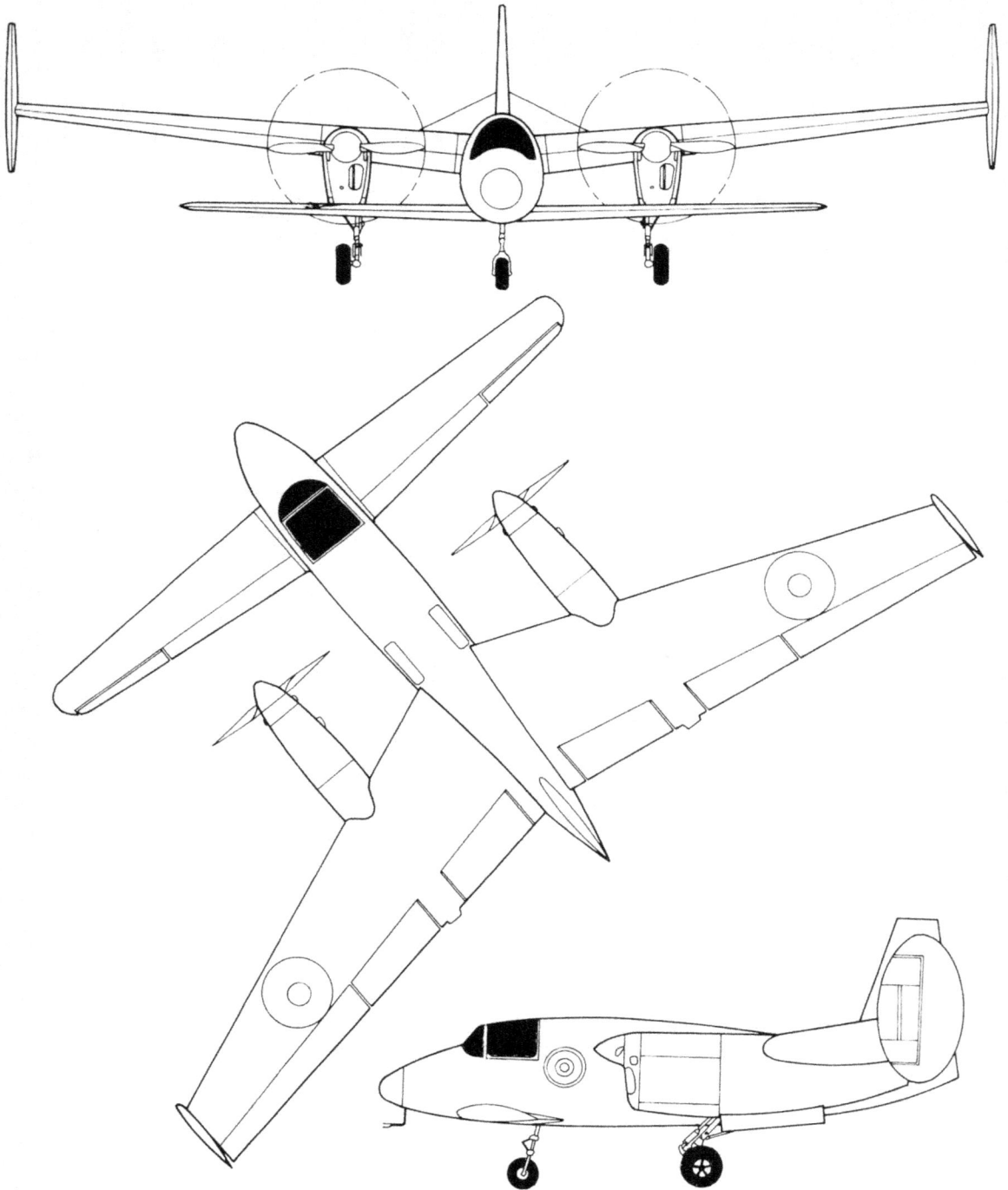

Fig. 2-39. Three view drawing of the Miles M.39B Libellula.

Fig. 2-40. The Japanese were also developing a canard at the end of WWI, in a vain attempt to produce an aircraft capable of warding off the high flying, heavily loaded B-29. A production goal of 1,800 Shindens was set for 1946, but WWII's end precluded manufacture of this beautiful design.

flap deployment of 10° was the maximum for landing.

It was recommended that the forward wing should be equipped with more powerful flaps (perhaps slotted instead of plain) to permit use of full aft flaps to achieve minimum landing speeds. Also, both sets of flaps should be operated simultaneously and under one control, and that the ailerons be modified to improve control at the lower landing speeds.

Unfortunately, the Ministry of Aircraft Production, under wartime pressures, discontinued any further developments of this intriguing design.

KYUSHU SHINDEN J7WI - JAPAN 1945

About the same time as the British were testing the Miles Libelulla, the Americans were developing the ill-starred XP-55, and the Germans were drawing their canard fighter and bomber, Japan

was developing a canard fighter—their Shinden or Magnificent Lightning. In a desperate attempt to combat the high flying, heavily gunned Boeing B-29's, the Japanese ordered the Shinden into production in 1945, with a production goal of 1,800 airplanes for 1946.

The Shinden was developed after flight testing of three wooden gliders—one of which was equipped with a 22 horsepower Semi four cylinder engine. It became a heavily armed, shore based high performance interceptor for protection of the home islands.

The Shinden featured a foreplane with double slotted flaps as elevators, a moderately swept aft wing with ailerons, and split flaps. Twin fins and rudders on the wing trailing edge, projecting both upward and downward, provided directional control. Long stalky landing gear, tricycle and retractable, combined with small wheels on the

63

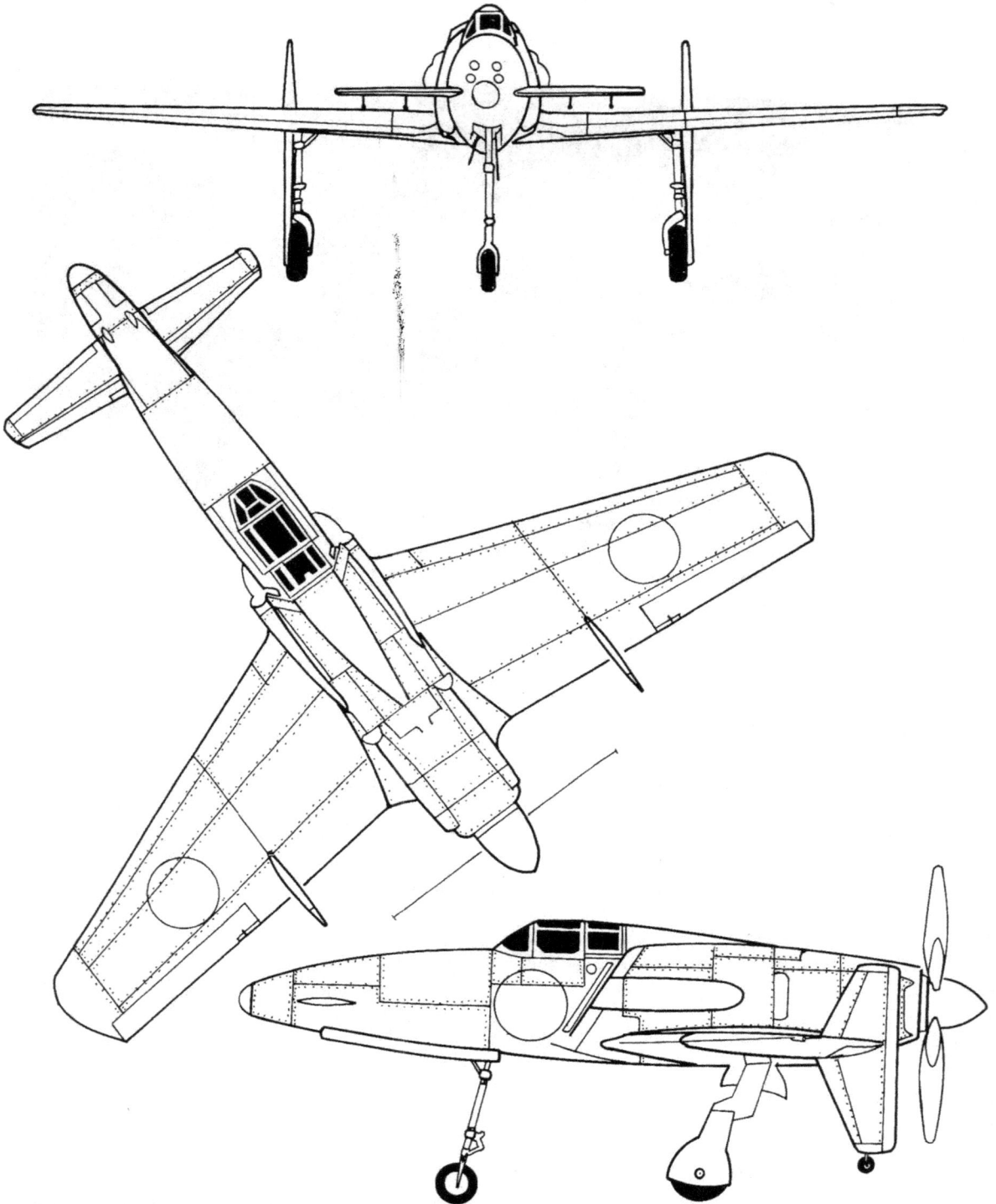

Fig. 2-41. Three view drawing of the Shinden.

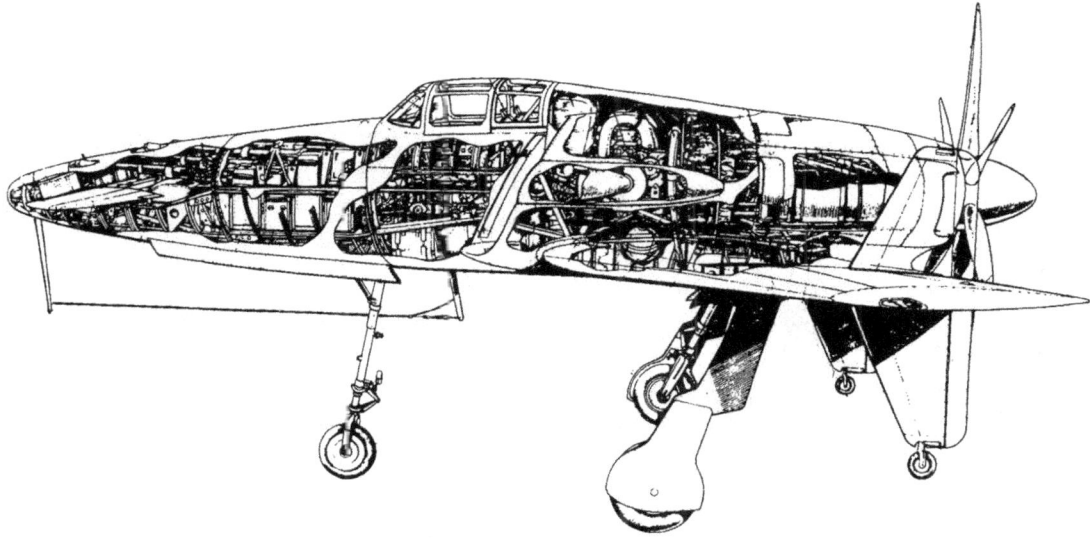

Fig. 2-42. Cutaway of the Shinden reveals its inner workings.

fin lower ends, served as undercarriage and also protected the propeller from ground contact.

The engine was a Mitsubishi, twin row, 18 cylinder air cooled unit of 2130 horsepower at sea level. It drove a six bladed constant speed pusher propeller through an extension shaft. Large engine cooling air intakes, along with carburetor air scoops, projected sideways behind the pilot's enclosure. Other cooling air intakes and outlets were incorporated further aft.

Four 30 mm cannon and one 7.9 mm machine gun in the fuselage nose gave it heavy fire power.

The Shinden was designed for a maximum speed of 468 mph at 28,000 feet, a takeoff speed of 126 mph, a landing speed of 117 mph. and a ceiling of 39,000 feet. Climb to 26,250 feet took 10 minutes 40 seconds. Range was 528 miles at 13,100 feet. Maximum weight was 11,526 pounds. Span was 35 feet and length 30.5 feet.

Two prototypes were built, and one was flown three times for a total of 45 minutes before hostilities terminated. Modifications needed were minor, a different cowling to eliminate prop flutter and automatic aileron tabs to overcome a strong righthand pull on takeoff. Cessation of hostilities resulted in the end of all work on the Shinden. One prototype is on display at the National Air and Space Museum, Silverhill, Maryland.

PONTIUS TANDEM WING BIPLANE - United States 1953

The "Pontius," designed by John Pontius, was a diminuative single-seat, tandem wing biplane, with tricycle landing gear. All controls were operated from the steering wheel.

The span was 20 feet, length 12 feet 6 inches, gross weight 720 pounds, and empty weight 453 pounds. Power was produced by a 36 horsepower Aeronca engine which provided 36 miles to the gallon.

The airplane obviously borrowed much from Mignet's Pou-du-Ciel, but instead of a tilting-forward wing, the Pontius employed a fixed wing solidly mounted on a streamlined cabin that provided 360° of visibility. The longitudinal pitch control was an aileron-like plain flap on the trailing edge of the forward wing that rotated downward for climb and upward for descent, just the opposite of the elevator on a rear-tailed airplane. These were actuated in the usual fashion by pushing and pulling on the control wheel. Rudder control was obtained by rotating the wheel, and like the Pou, banking in turns occurred as the air-

Fig. 2-43. American John Pontius built this interesting little tandem wing in 1953. Obviously inspired by Mignet's Pou, the Pontius featured a two-axis control system—pitch was controlled by an elevator on the forward wing, while the rudder yawed the aircraft, which in turn, skidded and rolled.

plane skidded sideways. Landings in crosswinds meant "crabbing" into the wind, until just before touchdown, and then sharply applying rudder to align the aircraft with the runway.

The pontius demonstrated both inherent stability and effective control during 18 hours of test flying over a period of nine months. The airplane could be flown "hands-off."

This aircraft was unusual in that the fuselage did not markedly change its pitch attitude as the elevators were actuated, either for climb or descent. When the wheel was pushed forward for descent, aerodynamic action prevented a nosedown pitch beyond a predetermined limit.

Pontius made use of two features in his design. First, the trailing edge of his foreplane acted as a plain flap—angled downward it induces a nosedown pitch, angled upward it induces a noseup pitch. (Ref. 24) Second, due to the large vertical gap between fore and aft planes, the wake from the front wing passes over the rear wing, leaving the latter in the downwash "below" the wake centerline. Wake and downwash angles vary in proportion to the lift being generated by the front plane—down elevator increases the lift, while up elevator reduces it. The change in downwash angles resulting from lift changes, acting on the aftplane, changes its lift to compensate—holding the

aircraft in a relatively level attitude. (Ref. No. 40 and 23). The tricycle landing gear reduced the need for landing flair or takeoff rotation.

The fate of the Pontius is dimmed by time. but it did not generate the popularity it deserved. It was stall free and safe, but the lack of positive three axis control was undoubtedly unpopular.

AVRO 730, HANDLEY-PAGE AND VICKERS LONG RANGE SUPERSONIC BOMBERS - GREAT BRITAIN 1955

These three jet powered aircraft designs were submitted to the British Air Ministry under specification RB-156T, as long range supersonic bombers.

All three were Delta wing canards. The AVRO had a rectangular foreplane and was powered by eight Armstrong Siddeley P.176 turbojets mounted in two-level pairs in nacelles near the wingtips. The Vickers had a delta foreplane and twelve RG-121 turbojets mounted below the wing. All three had huge central aft fins and rudders mounted above the fuselage.

Unfortunately, all these designs were to no avail as the result of the 1957 British Defence White paper. The interesting feature of all three is their close resemblance to the North American XB70 VALKYRIE in many respects.

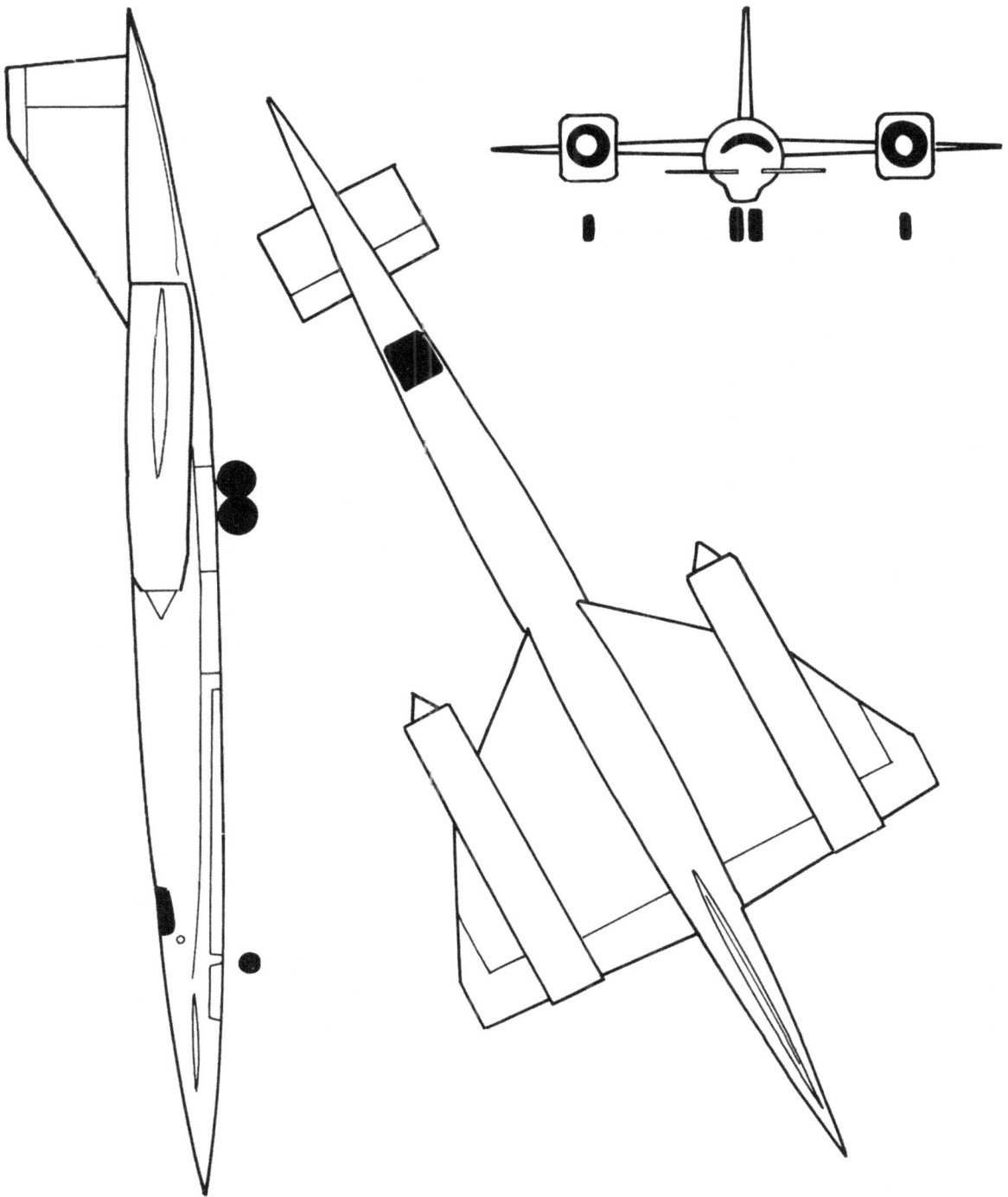

Fig. 2-44. Three view drawing of the AVRO 730 Lcng Range Supersonic Bomber of 1955.

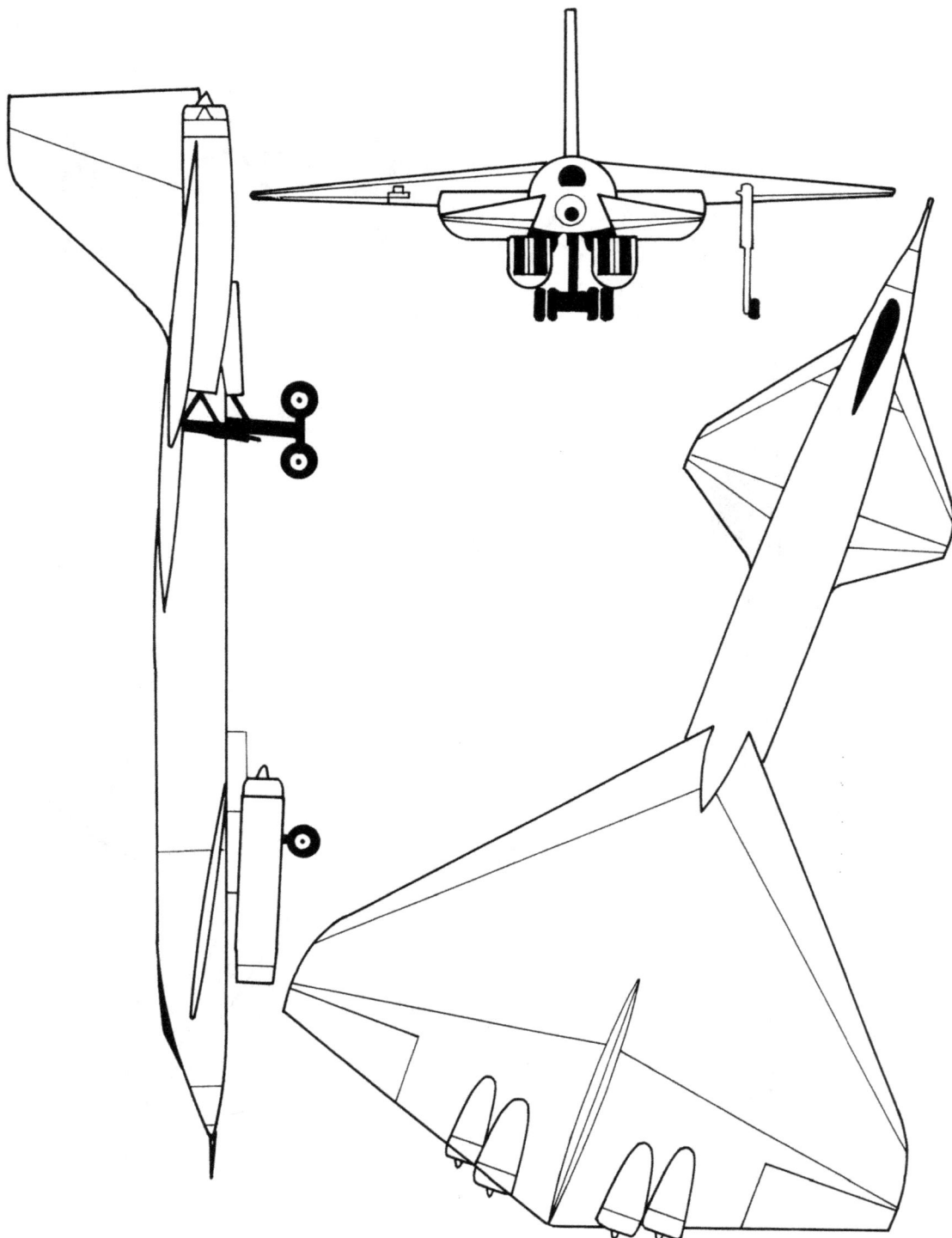

Fig. 2-45. Three view drawing of the Handley-Page Long Range Supersonic Bomber of 1955.

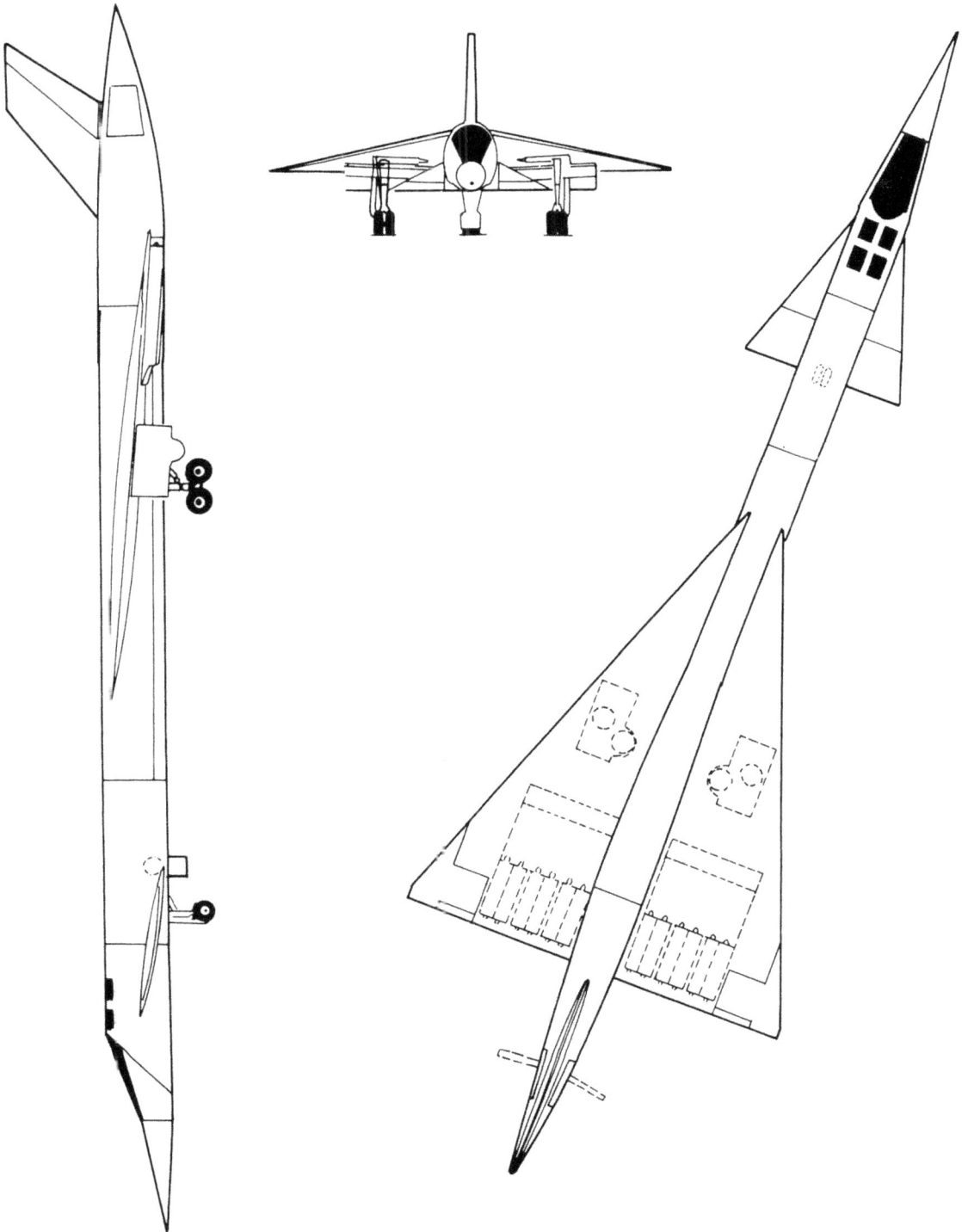

Fig. 2-46. Three view drawing of the Vickers Long Range Supersonic Bomber of 1955.

DESPRETZ JIDNEY JI3 "FLASH" - FRANCE 1965

The "Flash" was a tandem wing biplane, with tricycle gear, tractor engine, with fin and rudder carried on the fuselage extension aft of the rear upper wing.

Its engine was a 40 horsepower Persy LL four cylinder, partially cowled. The airplane was heavy at 880 pounds gross.

The wings were one piece, with full span, controllable, slotted, trailing edge surfaces. Those on the foreplane served as elevators and on the machine, the wing sections were both semi-symetrical.

The aircraft was stall proof, and easily controlled in all rolling motions, including a steep banking turns. It could fly slowly and land short with full "backstick"—that is to say "full down elevator" (nose up).

An improved version, the model JI-31 was designed which incorporated changes in wing shape, control (ailerons doubling as elevators) and span. Conventional tail wheel gear was incorporated, along with a fully cowled engine and enclosed cockpit. The subsequent history of this aircraft is unknown.

NORTH AMERICAN AVIATION XB-70 VALKYRIE — UNITED STATES 1964

This very advanced airplane was originally planned as a strategic bomber, according to specifications laid out by General Curtis LeMay, then head of the Strategic Air Command and Air Force Chief of Staff. It called for an unrefueled range of 6,000 miles, and a capacity to reach 1,100 mph in short bursts.

Fortunately, at this time, NASA researchers discovered the "compression lift" principle of supersonic flight. With a properly designed fuselage an aircraft could ride its own shockwave the same as a three-point hydroplane rises on and skims over the water, instead of plowing through it. It was now possible for the aircraft to cruise its whole mission in excess of the specification. During its trials, the XB-70 exceeded MACH 3, or three times the speed of sound, at an altitude of 70,000 feet, using boron-base high energy fuel.

The USAF foresaw a fleet of 150 of these planes by 1970 at a total cost of 19 billion dollars. Unfortunately, it fell victim to the fued between manned bomber advocates and those favoring ballistic missiles. Instead of quantity production, only two prototypes were to be built as experimental aircraft. The first XB-70 was rolled out in Los Angeles in May 1964.

In the nine intervening years, revolutionary solutions to some major problems were conceived. At supersonic speeds, despite 60 below zero outside air temperatures at cruising altitudes,

Fig. 2-47. The French Despretz JI-31 Flash tandem wing of 1965. Both wings featured full span, slotted trailing edge flaps.

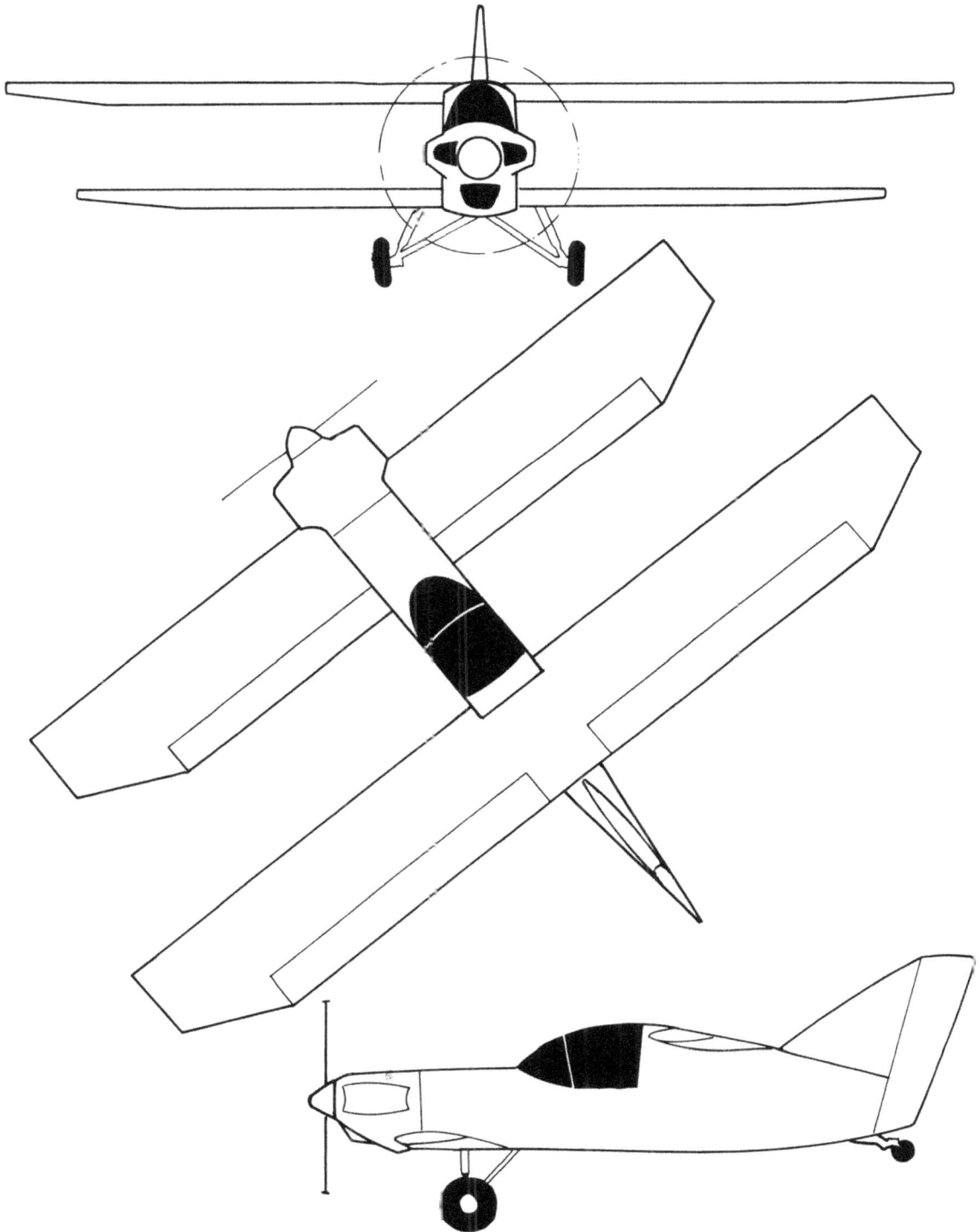

Fig. 2-48. Three view drawing of the Despretz Flash.

Fig. 2-49. The North American XB-70, was the largest, fastest canard ever made. The 269 ton aircraft could reach Mach 3.

air friction could generate skin temperature up to 600°. It was decided that the 150,000 pounds of jet fuel could serve as a heat-sink to cool the aircraft's skin. Such frictional heat could be absorbed during up to five hours of MACH 3 flight.

Other major technical advances were extensive use of titanium in the forward fuselage, and of three layer stainless steel honeycomb construction for up to 70% of the aircraft's weight. Self-contained ejection capsules, for survival if the crew was forced to bail out at high altitude and speed, were incorporated.

Aerodynamically, the XB-70 had a slender snake-like fuselage 170 feet in length, a relatively small foreplane, unswept, but with an elevator roughly half the area of the front wing. Aft wingspan was 115 feet, and of delta shape. From the front, this wing was thin, and had anhedral and twist. Twelve separate elevons occupied most of the wing's trailing edge. The outer wing rear cor-

ners folded down 25° for low level supersonic flight and 65° for high altitude MACH 3 cruise. Twin vertical fins and rudders provided directional control.

Massive power was provided by six General Electric J-93 turbojets of 25,000 pounds each, without afterburners, housed in a huge fairing, with two large air-intakes. This also contained the two sets of four wheel bogies, forming the main landing gear, as well as the two wheel steerable nosewheel.

The four man crew occupied a cabin just ahead of the foreplane, behind a peculiarly shaped nose. Forward visibility, on the landing approach at high angle of attack must have been poor, since the "droop-snoot," later incorporated in the supersonic Concorde air liner, was not used. Gross weight of this huge airplane was 538,000 pounds (269 tons!).

It is strongly suspected that the large foreplane

flaps were intended, when deployed, to hold the aircraft at the very high angle of attack required of delta wings at their maximum lift. (Many who have watched the Anglo-French "Concorde" landing, in real life or on TV will have noted the extremely high nose angle this delta winged aircraft assumes).

The XB-70's relatively small aft elevons seem to have been split in function: inner panels for high speed pitch control, and outer panels for roll.

Initial flight tests were "hairy." Hydraulic failures, and electrical shorts nearly caused what would have been disastrous crashes. Only superb piloting and ingenuity preventing disaster.

In June 1966 disaster struck. In a GE public relations sponsored demonstration flight of the second XB-70 and several fighter aircraft, all powered by General Electric engines, one of the fighters, an F-104, was caught in the right wing vortex. The F-104 bumped the tip with its tail, was rolled on its back by the powerful swirling vortex into the XB-70's twin vertical tails, shearing one off and damaging the other. The huge canard pitched sharply, rolled on its back and crashed. The pilot of the XB-70 parachuted to safety; unfortunately, his co-pilot and the F-104 pilot were both killed.

The first XB-70 continued test flying, providing information on sonic booms, engine and airplane performance, and other areas of SST research. It is now a museum piece, on display at the Airforce Museum, near Wright-Patterson Air Force Base, Dayton, Ohio.

It is baffling to the author to comprehend why the demonstrated high performance, delta-canard configuration is being ignored in the design of the current B-1 swept wing bomber, itself the subject of the "manned bomber" versus "ballistic missile" controversy.

4

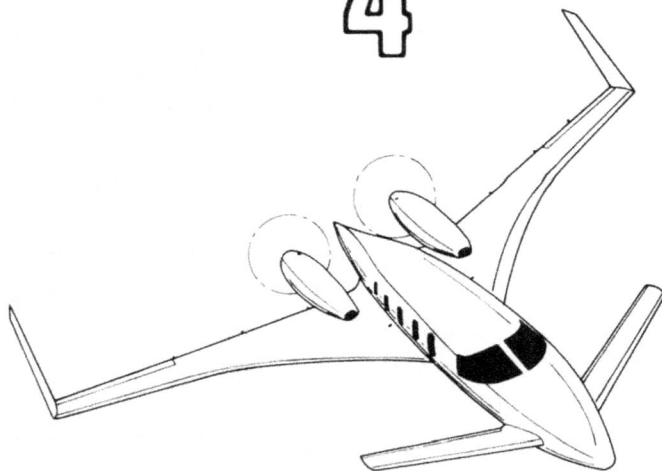

Hybrid Aircraft

"Hybrid" is the author's description for those airplanes of the 1907 to 1913 period that had horizontal control surfaces both in front as well as behind the main wings. Langley's "Aerodrome" of 1903 does not fit this category exactly: it has tandem monoplane wings, plus a tail that carried both horizontal and vertical surfaces. Its configuration was unique, and it is included in the "hybrid" category, as being the one most defining its design.

Landing gears for this class (except the Aerodrome, which had none) were both tricycle and taildragger—with either skid(s) or tail wheels.

Longitudinal or pitch control was centered in the front horizontal surface in some or combined with conventional elevators on the trailing edge of the rear horizontal surfaces, in others.

All had vertical surfaces to the rear with an occasional forward vertical surface.

With the exception of the Walden Monoplane all had biplane main wings. Langley's could also be called a "tandem wing" biplane.

Some, notably the Farman and Voisin aircraft, had aft horizontal surfaces that were cambered, as much as the main wing surfaces, indicating that, in flight, they produced lift. Of necessity the centre of these machines must have been behind the center of lift of the main planes.

The engines of that time were heavy for their horsepower output. The location of the pusher propellers, behind the main wings, and mounted directly on the engine out-put shaft, meant that the engine location was relatively fixed, necessitating an aft CG location. The Wrights overcame this problem by use of extension shafts, permitting a forward engine location with aft pusher propellers. Some of the designers utilized wing trailing edge cut-outs to permit both engine and propeller to be a little forward.

Ailerons mounted on the wing trailing edges, and hanging down when the aircraft was at rest, or composed of separate small wings located between upper and lower wing tips, provided lateral control. Engines varied from four cylinder in-line to Gnome 7 cylinder rotaries.

By 1912 the tractor engined airplane was becoming popular—and this spelled the end of the Hybrids, until 1983.

Fig. 2-51. The Langley tandem winged Aerodrome rests on its launching ways on a houseboat in the Potomac River, 1903. The aircraft did not fly, leaving the way open for the Wrights' success.

LANGLEY "AERODROME" - UNITED STATES 1903

Dr. Samuel Pierpoint Langley was noted as an astronomer, physicist and inventor. and third secretary of the Smithsonian Institution.

He conducted many experiments with a whirling arm device to develop wing sections, and in 1887 built many rubber powered models on which he tried his ideas. These proved too erratic. and Langley commenced a series of much larger models of 14 feet span. Power was provided by light, steam engines.

His Aerodrome No. 5 achieved outstanding results. Tandem monoplane wings spanned 13 feet 11 inches, and the machine weighed 26 pounds. A steam engine of one horsepower drove twin propellers, located aft of the front wing, through bevel gearing. Lateral stability was provided by a large dihedral, and was maintained longitudi-

nally by a four vaned tail, set at a negative angle, with its movement spring loaded.

Unmanned and launched from a catapult mounted on a house boat on May 6, 1896, it flew over the Potomac River in stable mounting flight. After a half mile, its fuel was consumed and it glided gently to the water. Dried out and refuelled, it was re-launched the same day for a second successful flight.

A second model, Aerodrome No. 6, with improvements in engine, propellers and wings, flew almost a mile in November 1896. This was a tremendous success for its time.

Arising out of the war with Spain, the U.S. Government recognized that an aeroplane, which could be guided on its course, was a much better perch for a military observer than a balloon, which when tethered had limited value, and when free, was at the mercy of prevailing winds. Accord-

75

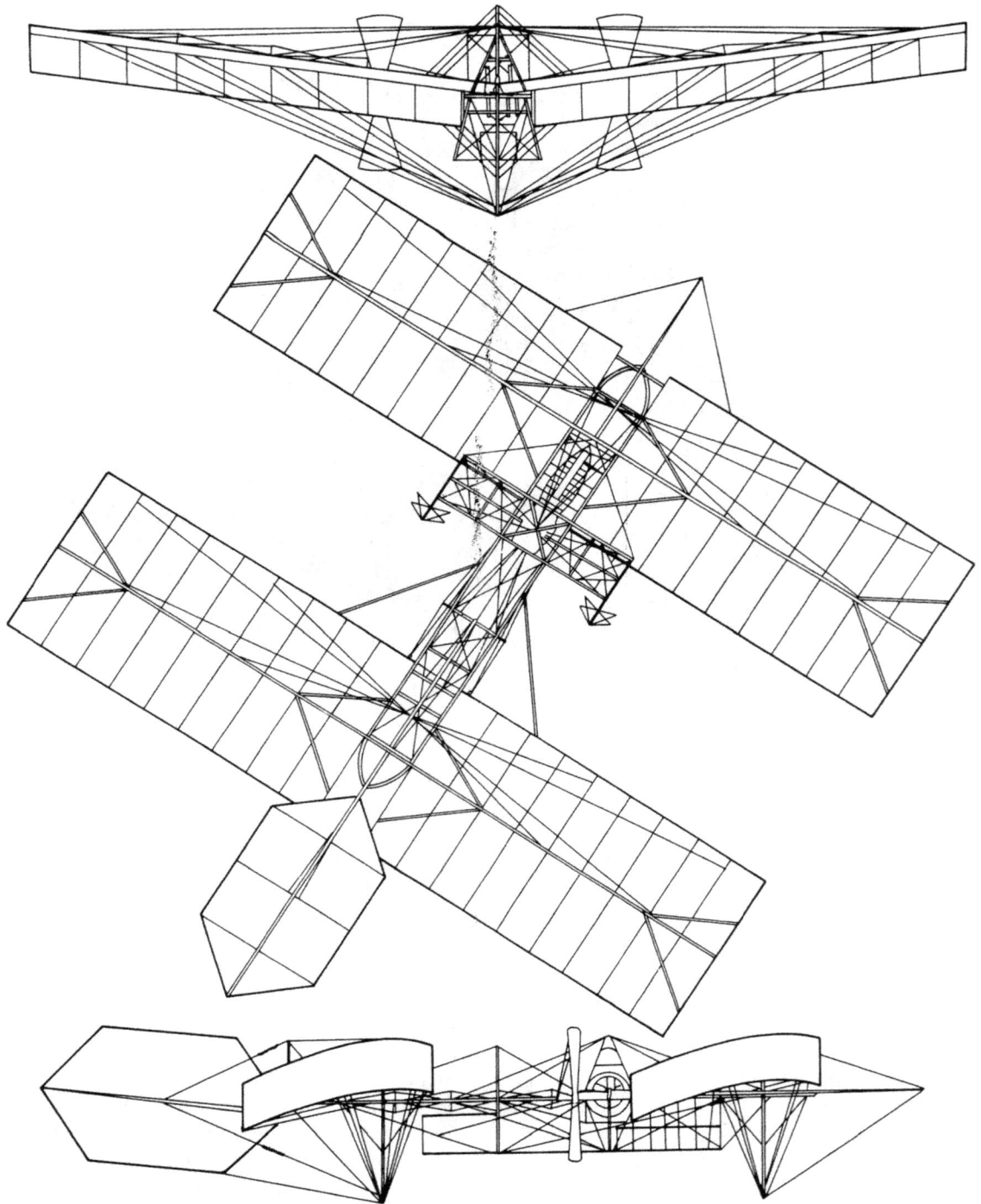

Fig. 2-52. Three-view drawing of the Langley Aerodrome.

ingly, Langley was granted $50,000 for construction of a man-carrying "Aerodrome."

The major problem was though: to be securing an engine of adequate horsepower. A Stephen M. Balzer built an air-cooled rotary engine under contract with Langley, which produced only eight horsepower instead of the required 12 horsepower. Mr. Charles M. Manly, Langley's Assistant, redesigned this engine. It produced a water cooled radial of 52.4 horsepower at 950 rpm, with a weight of 2.8 pounds per horsepower, which was quite remarkable for 1903.

A miniature airplane was constructed, gasoline powered, which flew very well on August 8, 1903. The full scale Aerodrome's design was based on this miniature, with a wingspan of 48 feet 5 inches,

length 52 feet 5 inches and weight of 750 pounds including Manly, the pilot.

This plane was launched by catapult from the houseboat roof, on October 7, 1903, but a pin on the launching car failed to release and the craft dove into the Potomac River. A second attempt on December 8, 1903, again ended in dismal failure. Langley, discouraged and out of funds, gave up. Public criticism of the U.S. Government for "wasting" funds on an impractical dream, and sarcastic comments in the newspapers are alleged to have hastened Langley's death, which occurred in 1906.

Restored to its original appearance, the "Aerodrome" was put on display in the National Air and Space Museum, in tribute to Langley.

Fig. 2-53. The successful 1/3-scale Langley Aerodrome of 1896 flew for almost a mile landing intact in the Potomac River.

One unanswered question is just what type of controls incorporated in Langley's Aerodrome. Other than the small rudder just ahead of and below the aft wing for directional control, nothing else seems to have existed, although the models proved the configuration stable.

VOISIN-FARMAN I - FRANCE 1907

France was the country to establish limited volume production of aircraft as an industry. The Voisin Brothers. Gabriel and Charles, set the lead in setting up an assembly line of boxkite type biplanes for sale to anyone who had adequate funds and was daring enough to fly them.

One such individual was Henry Farman, an Englishman who lived in France and became a naturalized Frenchman. His airplane, the Voisin-Farman I, was a biplane of 33 feet - 5.5 inches span, length 34 feet - 5 inches, and weighed 1,150 pounds. The engine was an Antoinette eight cylinder (two banks of four cylinders now known in automobile parlance as a V-8). The main landing gear was a well sprung "trailing link" design, located just below the pilot. Two smaller wheels supported the tail. The fuselage was partly enclosed and carried the forward wing at its extremity.

Pitch control was provided by tilting the front plane about a lateral axis. Lateral control was some form of wing warping. No rudders were used, although three vertical fins were provided by fabric covering between the interplane struts of the biplane tail. The moment arm between wing and aft tail was very generous for stability.

After test flying, Farman made extensive modifications incorporating his own ideas. He installed "balancing flaps" (ailerons) on the trailing edges of all four wing tips and several other changes.

The modified airplane flew very successfully, at Issy, France, on October 7, 1907. On October 26 it flew 2,530 feet in 52.6 seconds. On November 5, 1907 it made its first turn in flight and broke the "one kilometer in one minute" barrier (38 mph).

It was not until Wilbur Wright's 1908 demonstration flights that France in particular, and Europe as a whole, realized how much further these two brothers had advanced aviation than was true in Europe.

CURTIS JUNE BUG - UNITED STATES 1908

This was an arial Experiment Association design that launched the career of one of the most famous Aircraft builders, Glen Hammond Curtis. It preceeded McCurdy's Silver Dart, and incorporated festures that later appeared in the Dart.

The engine was a Curtis air cooled V-8 of 40 horsepower. Its length was 27 feet - 6 inches, span 42 feet - 6 inches, wing area 370 square feet, and gross weight 615 pounds. Speed was 39 mph.

Like other Hybrids, the engine was located near the main wing trailing edge, driving a pusher propeller. Pitch control consisted of a small monoplane forward wing, controllable about a lateral axis: lateral control was by triangular wing tip ailerons. The horizontal tailplane was carried well aft and incorporated rudders.

The landing gear was tricycle. The airplane made over thirty successful takeoffs in late June 1908. On July 4, 1908 it made the first *official* flight in the United States to cover a distance of more than a kilometer and was awarded a prize offered by the Scientific American. The flight covered 6,000 feet in a time of 1 minute 42.5 seconds. Obviously the Wright brothers' achievements were no *"official."*

Curtis went on to achieve greatness on his own. The name Curtis is almost synominous with aviation in the United States.

DEHAVILLAND BIPLANE NO. 1 - GREAT BRITAIN 1909

Geoffrey DeHavilland was the founder of a firm that became known world-wide for its airplanes.

His famous Moths—so called because the vertical tail surfaces had an outline like that of a moth's wing—have been used to train more pilots than almost any other airplane. The Gypsy Moth two-seater of the thirties (your author took his first flying lessons in one), the Tiger Moth, similar to the Gypsy except for swept back biplane wings, along with the Puss Moth and Fox Moth, are all famous DeHavilland airplanes.

DeHavilland of Canada, now a crown corporation, is known for its Beaver, Otter, Twin Otter, Buffalo, and current commercial airplanes, the Dash 7 and Dash 8. The DH Mosquito, an all

JUNE BUG
1908
CURTISS EXH. Nº

1440

Copyright 1908 H M Benner

Fig. 2-54. The 1908 Curtiss June Bug was the first airplane to *officially* fly further than a kilometer, and was awarded the *Scientific American* trophy for the feat.

wood airplane, was one of the most versatile aircraft of World War II, and the DH Vampire jet engined fighter first flew in 1943.

Despite this great success, DeHavilland's first biplane was a dismal flop. The airplane had a wing span of 36 feet, a length of 29 feet, a wing area 408 square feet, and it weighed 550 pounds. Its engine was a DeHavilland four cylinder, horizontally opposed water cooled unit producing 45 horsepower, turning two shaft mounted pusher propellers through shafting and gearing, not unlike the Wright brothers arrangement.

The landing gear had the main wheel located below and ahead of the pilot. The steerable tail wheel was located aft on struts that held the aircraft in level attitude at rest. A front wheel was also provided to prevent nosing over. The biplane wings had ailerons on the top surface. These were

small, and typically drooped when at rest.

The foreplane controlled pitch and a vertical rudder provided directional control. The aft horizontal stabilizer was cambered and fixed.

The plane crashed during its only test flight at Seven Barrows near Newbury, England, in December 1909 when one wing collapsed after being airborne for only a few seconds.

A somewhat cruel humorist of the day was heard to say "if a pigeon is released between the wings and it escapes to the open air, a bracing wire has been missed." This does not imply that the crash of DeHavilland's airplane was due to a missing wire, but it does point out the complexity of the inter-plane struts and bracing.

DeHavilland salvaged the engine, of his own design, from the wreckage and installed it in his aircraft.

79

CURTIS GOLDEN FLYER - UNITED STATES 1909

Glenn Curtis, now on his own, built the Golden Flyer, in partnership with a Mr. Augustus M. Herring, for the New York Aeronautical Society, which paid $5,000 for the plane. With it, Curtis won the *Scientific American* prize for a non-stop flight of 25 miles.

The aircraft was 28 feet - 9 inches in span, with a length of 33 feet - 6 inches. Wing area was 258 square feet and it weighed 550 pounds. Its speed was 45 mph. The main wings were of equal span, arranged one above the other, with a gap between roughly equal to their chord. The center-section of both wings was recessed to provide propeller clearance. The ailerons were small uncambered surfaces placed midway between the wings, and extended beyond the wing tips some three feet.

A small biplane forward wing, also flat sur-faced, pivoted on a lateral axis and provided pitch control. Directional control was via a rudder mounted well aft, above and below a fixed horizontal tail plane. The landing gear was tricycle; the main wheels were located below the engine and the third wheel was located forward, but was not steerable.

The engine was a Curtis design, four cylinder in-line water cooled of 25 horsepower. It was mounted with the crankshaft half-way between the upper and lower wings, and swung a six foot pusher propeller. Subsequently, 50 horsepower Curtis four cylinder, and 100 horsepower V-8 Curtis water cooled engines were installed.

The control arrangements were typical of the hybrids at that time. The forward wings were operated by a push-pull on the wheel; rotating the wheel operated the rudder. An aileron Yoke, against which the pilot pushed his shoulders, pro-

Fig. 2-55. Glen Curtiss built his Golden Flyer in 1909, and won the *Scientific American* trophy for a non-stop flight of 25 miles.

Fig. 2-56. In 1909, Henry Farman broke away from the Voison brothers, and started building his own airplanes. This one, the Henry Farman III, was Europe's most popular between 1909 and 1911.

vided roll control. The ailerons did not droop when the aircraft was at rest. The pilot's seat was ahead of and below the engine—a dangerous spot in a crash. The plane was strong, fairly fast for the times, and easy to handle, largely attributable to the ailerons between the two wings.

A development of the Golden Flyer, the Riems Racer with reduced wing area, powered by a 51 horsepower V-8, was built and won the Gordon Bennett Trophy on August 28, 1909 and the Prix de la Vitesse the following day. Later Curtis Aircraft made takeoffs and landings from the deck of a warship on November 14, 1910, and again on January 18, 1911 piloted by Eugene Ely.

Curtis went on to develop flying boats and amphibians, capable of operating from both land and water.

HENRY FARMAN III - FRANCE 1909

Farman had a falling-out with the Voisin Bro-

thers when the latter sold the aircraft he had ordered to someone else. Infuriated, Farman established a small factory in which he built a new airplane of his own design. The result was the Herry Farman III, which was the most popular airplane between 1909 and 1911.

At the Reims Aviation Rally in 1909, this aircraft won the "Grand Prix" for flying 112 miles in three hours, four minutes and 56 seconds, the "Prix Des Passengers" for flying with two passengers, plus the pilot, and second place in the "Prix de L'Altitude" for flying all of 361 feet up.

It originally flew on April 6, 1909 powered by a four cylinder 50 horsepower Vivinus engine— which was subsequently replaced by a 50 horsepower Gnome rotary. Further improved and modified, the Farman became the most widely used aircraft prior to World War I.

The 1910 Farman spanned 34 feet 10 inches, was about 38 feet long, and was powered by the

50 horsepower Gnome rotary. One machine, flown by Paulhan, won the Daily Mail prize of 10,000 pounds for the first flight between London and Manchester, England.

Some versions had equal span main wings, others had the lower wing shortened by one bay. The rotary engine with the propeller mounted on its front face was located low, just above the lower wing, so that the propeller operated in a cut out portion of the lower trailing edge. Pitch control came from a cambered forward monoplane wing, mounted on lateral pivots so that its angle of attack could be changed. This surface was connected by cables to elevators on the trailing edge of the upper wing of the biplane horizontal tail. Twin rudders, at the extremities of the tail, provided directional control, along with large ailerons on the upper wings for roll control. They drooped at rest.

The landing gear consisted of two main wheels just below the pilot, sprung on rubber bungee cord, with aft wheels, and skids to provide braking action on landing. This gear permitted the airplane to be flaired for landing, at close to the wings' maximum lift and slowest speed.

Controls consisted of a stick pivoted on a universal joint at the right of the pilot's seat. Fore and aft motion of this stick operated both foreplane and aft elevators in concert. Lateral action pulled down the balancing flap (aileron) on the side in the direction of stick action. These "flaps" trailed horizontally in the airstream in flight, so that only one, not both, were operated. The rudders were controlled from a rudder bar which also served as a foot rest. The passenger(s) occupied a seat behind and slightly above that of the pilot.

Farman went on to develop pusher powered reconnaisance aircraft in World War I, and a large twin engined biplane bomber in 1918. It was modified for civil use as a transport in Europe for a period of ten years, notably on the Paris-London route.

MARTIN'S FIRST BIPLANE -
UNITED STATES 1909

Glenn L. Martin was another aviation pioneer who went on to found his own manufacturing organization. It became famous for many flying boat designs including the four engined "China Clipper," the PBM "Mariner," and during World War II for the twin engined B-26 Martin "Marauder," medium bomber capable of carrying a large bomb load. Like many other famous airplane manufacturers founded by aviation pioneers, the Martin Company has disappeared into mergers, consolidations, etc.

Martin's first airplane appeared in 1909, powered by a 12 horsepower Ford engine. It was a biplane not unlike Farman's, with a forward monoplane wing, controlled in pitch and connected to elevators on the aft stabilizer trailing edge. A single rudder provided directional control, and the airplane had interplane ailerons similar to those of Curtis. Tricycle landing gear was provided. Controls were a hand wheel for pitch and roll, plus a rudder bar. Little else is known concerning this obscure airplane.

SHORT BIPLANE NO. 3 -
GREAT BRITAIN 1910

The Short Brothers: Horace, Eustace and Oswald, had considerable success with earlier Wright type machines, but their Short Biplane No. 3 was an unusual example of technological regression. It was designed, built and put on a display at the Olympia Aero Show of 1910, but repeated attempts to effect takeoff were unsuccessful.

The original undercarriage was Wright-like skids, later modified to incorporate four wheels in a vain attempt to fly. The plane, while a hybrid, had both its pitch control, a biplane forewing, and its rudder up front. The aft surfaces were fixed. Roll control was effected by small interplane ailerons extending beyond the main plane wing tips. The engine was a Green four cylinder in-line water cooled 35 horsepower unit.

The wings spanned 35 feet - 2 inches, and it was 31 feet long. The wing area, at 282 square feet, was comparatively low, and the aircraft weighed 860 pounds loaded. The low wing area plus the weight resulted in heavy wing loading, which may have been its undoing.

The Short Brothers founded another aviation dynasty, best known for their long range, four engine flying boats: the Canopus, Sunderland, and Shetland. Today they are manufacturing high wing, twin turboprop powered commercial aircraft for the commuter airlines, many of which are used in the United States.

WALDEN MONOPLANE - UNITED STATES 1910

Doctor Henry W. Walden built and flew the first American monoplane in 1909, and made 12 airplanes between 1908 and 1913. He accumulated over 600 flying hours before 1914.

The monoplane was powered by a three cylinder, 20 horsepower air cooled Italian Anzani engine. The monoplane wing was selected because of the believed aerodynamic interference between biplane wings. The machine had a monoplane forward wing pivoted laterally for pitch control, a monoplane horizontal tail surface equipped with trailing edge elevators, and a rudder. The landing gear was tricycle. Farman-type drooping ailerons provided roll control. Wingspan was 26 feet with

165 square feet of wing area. The propeller was six feet in diameter and 4.5 feet pitch.

During his aviation career, Walden became quite a showman, was feted widely, and crashed on more than one occasion. He gave up active participation in aviation after seeing others killed in his aircraft. Walden continued his career as a dentist, and is known more for his many patients and medical research.

CODY MICHELIN CUP BIPLANE - GREAT BRITAIN 1910

Samuel Franklin Cody was an American who made a considerable contribution to British Aeronautical Research.

In 1910 he built a biplane to compete for the first Michelin Cup. This aircraft was powered originally by a 60 horsepower Green that was replaced by an ENV of the same power. In it, Cody set a British record by flying 94.5 miles in two

Fig. 2-57. Glenn L. Martin's first aircraft was a hybrid, built in 1909.

83

Fig. 2-58. America's first monoplane was a hybrid, designed and built by Henry W. Walden, a dentist.

Fig. 2-59. The 1910 Cody hybrid won Britain's Michelin Cup for flying over 185 miles in less than five hours.

hours and 24 minutes. On December 31, 1910 he won the Michelin Cup covering 185.46 miles in four hours and 47 minutes.

His aircraft had an unusually large foreplane; the center of gravity was just ahead of the main wings. A small fixed horizontal tail plane was mounted behind the wing. The engine was located near the wing leading edge, driving the propeller via an extension shaft. The pilot sat ahead of the engine. A large, aft vertical surface provided directional control and the ailerons were located between the wings and extended slightly beyond the tips. The foreplane provided pitch control. Landing gear was a form of tricycle, with an added skid below the propeller.

Cody's further developments are obscure.

5

Water Canards And Hybrids

The early experimenters, Fabre, Curtis, Martin etc., laid the foundation for waterbased airplanes that culminated in the Boeing 314 Clipper, Vought-Sikorsky VS-44a, Martin-Mars JRM and the Saunders-Roe Princess, the largest seaplane ever built.

Today only a handful of companies are building seaplanes, amphibians, or floats to be used with normal airplanes. The economy, long-range and speed of the modern jet airplane, plus the airport facilities needed for commercial aviation, have rendered the waterbased flying boat impractical.

FABRE HYDRO AVION - FRANCE 1910
(The First Seaplane)

Many experimenters had endeavoured to develop an aircraft capable of taking off and landing on water, including such famous names as Kress, Voisin, Archdeacon and Blériot. It fell to Henri Fabre, an engineer and naval architect, to succeed first.

His seaplane was a canard of weird and wonderful design, reminiscent of the 1910 Valkyrie. Two long beams connected fore and aft wings. The main wing had an unusual latice work spar on the top surface, certainly not the best arrangement aerodynamically, but very strong. Biplane front wings, the lower fixed, the upper moveable, also with the unusual spars, provided both lift and pitch control.

Early versions had two vertical rudder surfaces mounted on the upper forward wing. A long lever, directly connected, provided pitch via push-pull control. When moved right or left, the lever moved forward rudders for directional control, which was found unsatisfactory and Fabre reverted to rear rudder control. A metal plow seat was mounted well forward, on which Fabre sat, with his feet on pedals that controlled the wing warping.

Three flat bottomed, cambered top floats supported this intriguing canard on the water. Originally, the front float could be turned for maneuvering, by the same lever that controlled the front rudders. Later modifications placed small water rudders to the rear of the aft floats. The secret of successful takeoffs was the inclined

Fig. 2-60. The world's first successful seaplane was built and flown by Henri Fabre of France, in 1910. It was a true canard.

(10°) flat bottoms, which permitted them to plane on the water surface. Further modifications were the addition of below-wing vertical surfaces to supplement the large rectangular vertical surfaces just ahead of the engine.

The engine was a 50 horsepower seven cylinder Gnome rotary, with a pusher propeller, mounted above the upper longitudinal beam. The propeller was of surprisingly large diameter and blade area for the size of the engine, and rotated at 1,100 rpm. The engine location must have required the aircraft to be backed up to a dock so that the propeller could be swung. This probably justified the backward projection at the bottom of the aft vertical surface, to keep the propeller from hitting the dock. The Gnome rotary engine could not be throttled to idle. Taxiing and landing approaches were achieved by interrupting the ignition, the mass of the rotating engine and propeller kept the

unit turning while the ignition was off. A landing approach had a typical rotary sound—short bursts of power interrupted by periods when only the "wind-in-the-rigging" was heard.

The wing covering was deeply scalloped between the ribs, which projected beyond the covering. It could be detached from the ribs and furled forward to the leading edge spar, similar to storing a sailboat sail. This was, no doubt, due to Fabre's naval background. Total lifting surface was 234 square feet and the aircraft had an all-up weight of 836 pounds, and a wing loading of 3.5 pounds per square foot.

Fabre, who had never flown before (like most of the inventive and daring aviation pioneers) took-off in his creation on March 28, 1910 from La Mede Harbour near Marseilles, France, for a flight of 1,600 feet at a height of six to eight feet above the water. Later that week, he flew 3½ miles

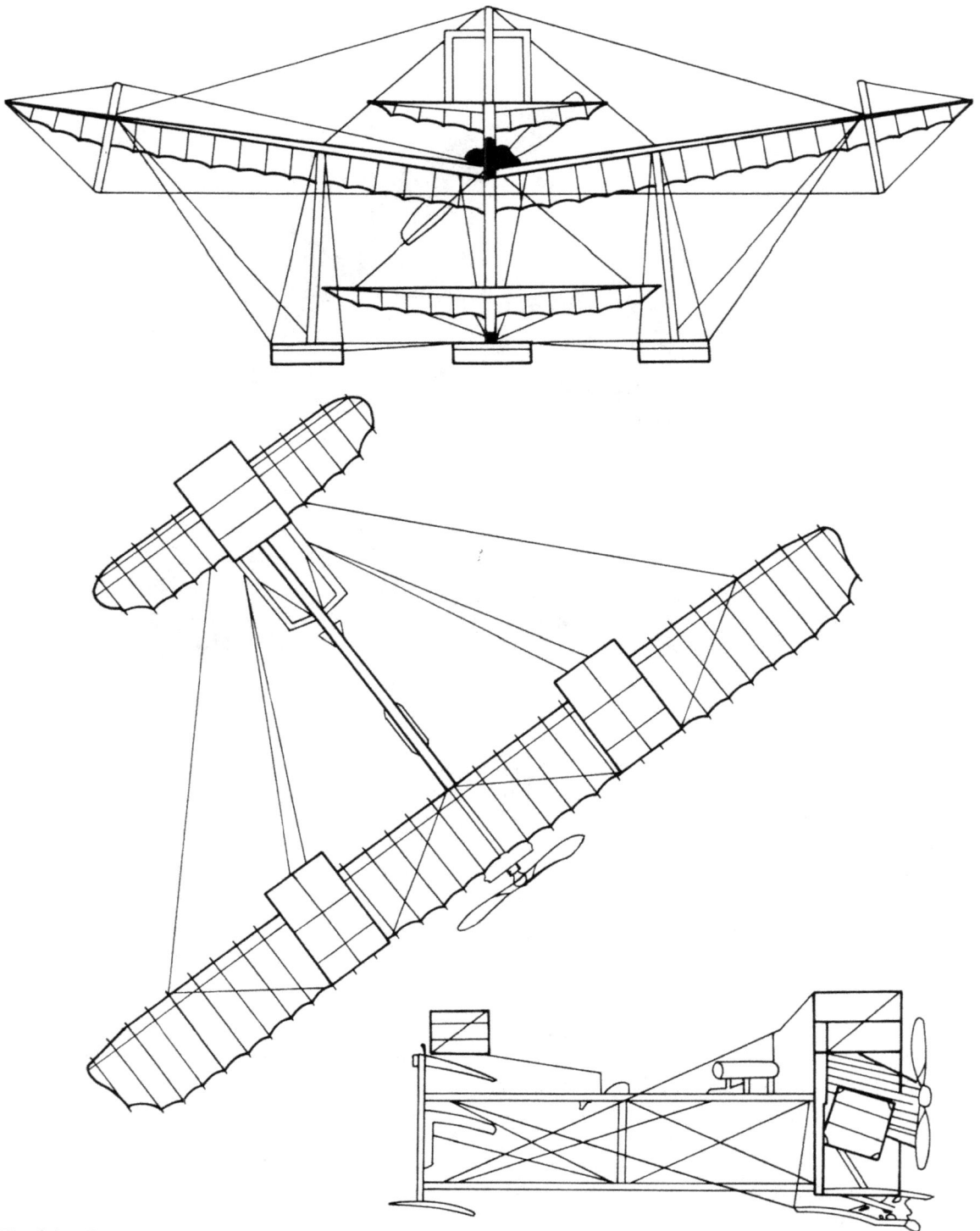

Fig. 2-61. Three-view drawing of the Fabre Hydro Avion.

Fig. 2-62. Henri Fabre preparing for another flight in his Hydro Avion. He sat halfway between the main wing and canard.

at an altitude of 40 feet. On May 18 he flew two miles at 75 feet. Subsequently he crashed the plane, but considering his inexperience, his achievements were sensational.

While it was rebuilt and flown again, it was wrecked in a gale. Restored, the Hydro Avion is on display at the Marseille Airport as of December 1982. Mr. Fabre lived to celebrate his 100th birthday in 1982. While he made floats for the Voisins, history does not reveal any further developments of his in aviation.

CURTIS AI AMPHIBIAN - UNITED STATES 1911

Glenn Curtis, famous for his Golden Flyer hybrid, did more to pioneer water aircraft in the United States, than any other individual. While he began testing a seaplane version of his June Bug in 1908, his first successful takeoff took place January 26, 1911 at San Diego, California.

This plane, and others on which he installed wheels as well as floats, resembled his Golden Flyer. The amphibian was called the "Triad" and proved practical. All had a central float and two small flotation units under the wing tips for lateral stability. The forward plane, instead of being supported by struts, was mounted on the central float upper surface. A Curtis water cooled V-8 was installed in the A-I. The aircraft had a span of 37 feet; was 27 feet - 8 inches long, and had 286 square feet of wing area. It weighed 1,575 pounds, for a wing loading of 5.5 pounds per square foot. Its speed was 65 mph.

Controls were similar to the Golden Flyer. Pitch was controlled by both foreplane and aft elevators. The wheel controlled the rudder when rotated and the ailerons were actuated by pushing on a yoke with the shoulders.

Curtis and the U.S. Navy cooperated in naval aircraft developments. Catapults for deck launching were devised and many record flights were made during this period.

VOISIN CANARD WATERPLANE - FRANCE 1912

This plane was patterned somewhat after Fabre's formula, and was a biplane canard. The foreplane was a monoplane structure controllable in pitch. The biplane aft wings were strut braced and carried large vertical surface, aft of the wings and outboard of the propeller, for directional control.

The fuselage was unusual in that it was fully enclosed and seated two. Fabre-type floats were used in a "tricycle" configuration — one small float up front and two larger ones aft.

The engine was a Gnome Rotary of 80 horsepower. Wing span was 44 feet - 3.5 inches, it was 36 feet long, and had a wing area 376 square feet. Its loaded weight was 1,212 pounds for a wing loading of 3.22 pounds per square foot. Speed was 62 mph. It should be noted that as speed increased, so did wing loading.

Voisin Freres did not pursue the development of water based aircraft.

The "tricycle" float arrangement devised by Henri Fabre did not survive but later float planes adapted two large front floats, with a smaller float below the tail surfaces. This arrangement duplicated that of land planes of the day; it permitted the aircraft to flair for increased angle of attack and the highest lift, while providing minimum takeoff and landing speeds.

By 1925 the tail float disappeared, being replaced by a stepped elongation of the two forward floats. This is still being used today.

CURTIS FLYING BOAT - UNITED STATES 1912

The first Curtis Flying Boat flew January 12, 1912. The totally enclosed hull, except for the pilot's cockpit, supported the front horizontal plane, the wings, power unit, and tail surfaces. It

Fig. 2-63. The Curtiss A-1 was the United States' first successful seaplane. Photo shows it in a low speed taxi.

Fig. 2-64. Curtiss' first real flying boat had its maiden flight in January, 1912. It featured an enclosed central hull, and canard elevator coupled to a rear elevator with horizontal stabilizer.

was the last of the Curtis hybrids. On later models, he dispensed with the foreplane; its low position on the bow must have presented considerable risk of damage.

Curtis persisted with the pusher engine and propeller location which had major advantages, for docking, boarding and unboarding passengers. It was aerodynamically more efficient, and provided slip stream pressure on the aft control surfaces. Best of all, neither pilot nor passengers had to sit in the noisy, dirty blast.

The original hull had a flat level bottom from the tail to the turn-up for the bow. This flat surface at the rear prevented the aircraft from "rotating" as it planed, so that the increased angle of attack could lift the aircraft off the water. Curtis then angled the rear under surface upward from a point just below the center of gravity. This led to the stepped hulls and floats used by all water planes to this day, where the under surface behind the step is angled upward.

Section III

Review Of Modern Canard And Tandem Wing Aircraft

6

Fighter Canard

The following pages contain a review of the canards and tandem wing airplanes, currently or recently flying. Included are homebuilt light airplanes, ARVs, ultralights, gliders and man-powered or solar/man-powered super light aircraft.

SAAB-SCANIA VIGGEN NO. 37 - SWEDEN 1967

The Viggen is basically a flying platform, adaptable to a wide variety of mission requirements. It is a MACH 2 STOL (Short Takeoff and Landing) aircraft—a complete weapons system, within the framework of Sweden's defense. Designed to operate from dispersed, wartime, road-type STOL runways, the Viggen can takeoff and land in 1,635 feet, truly remarkable for a machine capable of MACH 2 speeds.

Highly maneuverable at all speeds, the aircraft can climb from takeoff to 33,000 feet in less than 100 seconds. It is safe and easy to fly. Swedish Air Force student pilots are flying this jet after no more than 185 hours total flight training. Its avi-

onics include a central digital computer and heads-up-display (HUD). This Viggen system has great growth potential, and advanced versions will remain in production for some time to come.

The airplane is used in many variations for widely different roles. The AJ-37 is a fighter bomber—the JA-37 is a fighter interceptor—the SF-37 for strike reconnaisance over land—and the SH-37 for strike—sea surveillance duties. The SK-37 Viggen is a two seat trainer that is combat adaptable.

The JA-37 has an internally mounted 30 mm Oerlikon KCA cannon. On other versions, armament, missiles, bombs, and long range fuel tanks are externally mounted.

The Viggen is the first genuine modern canard military airplane to enter series production. The basic configuration is that of a Double Delta. The foreplane is fixed with a trimmable flap set ahead of and slightly higher than the Delta aftplane and at a noticeably higher angle of incidence than the latter. The main wing trailing edge contains the high speed flight controls—elevators inboard and

Fig. 3-1. The Saab-Scania Viggen is a Mach 2 STOL fighter. It originally inspired Burt Rutan's Vari-Viggen amateur-built airplane.

ailerons outboard. A large central fin and rudder provide directional control. The foreplane is highly loaded, so the aircraft is naturally stable, longitudinally.

The Viggen gains much from the favorable interaction between the close coupled foreplane and wing; the deployed flap of the former forcing the main wing to a very high angle of attack, without stalling—a capability characteristic of Delta wings. Non-flare landings are flown—the approach being made at high angle of attack all the way to ground contact, not unlike carrier landings.

The landing roll is shortened by thrust reversing and operating brakes on the main landing wheels, which are arranged in tandem pairs to retract into the thin wings. The nose wheels are a side by side pair. Visibility is superb.

Noticeable on takeoff, are the lowered foreplane slotted flaps and the raised main wing trail-ing edge elevators. All controls are power-boosted. As mentioned, the Viggen is capable of MACH 2 at altitude and MACH 1.1 as low as 330 feet above ground level.

Thrust is provided by a Pratt and Whitney Turbo Fan Engine, built under license by Volvo, and fitted with a Swedish built afterburner. This engine provides economical cruise.

Span of the Viggen is 34 feet - 9 inches and length is 53 feet - 5 inches. The main fin folds down to reduce overall height to 13 feet to fit under-ground hangar clearance. Takeoff weight for the AJ-37 is 35,275 pounds.

Sweden plans to put a total of 350 SAAB Viggens of different variations into service.

A follow-on design, the SAAB JAS-39 "Gripen," multi-role combat aircraft is now under full scale development for the Swedish Air Force. It is sim-ilar in configuration to the Viggen.

94

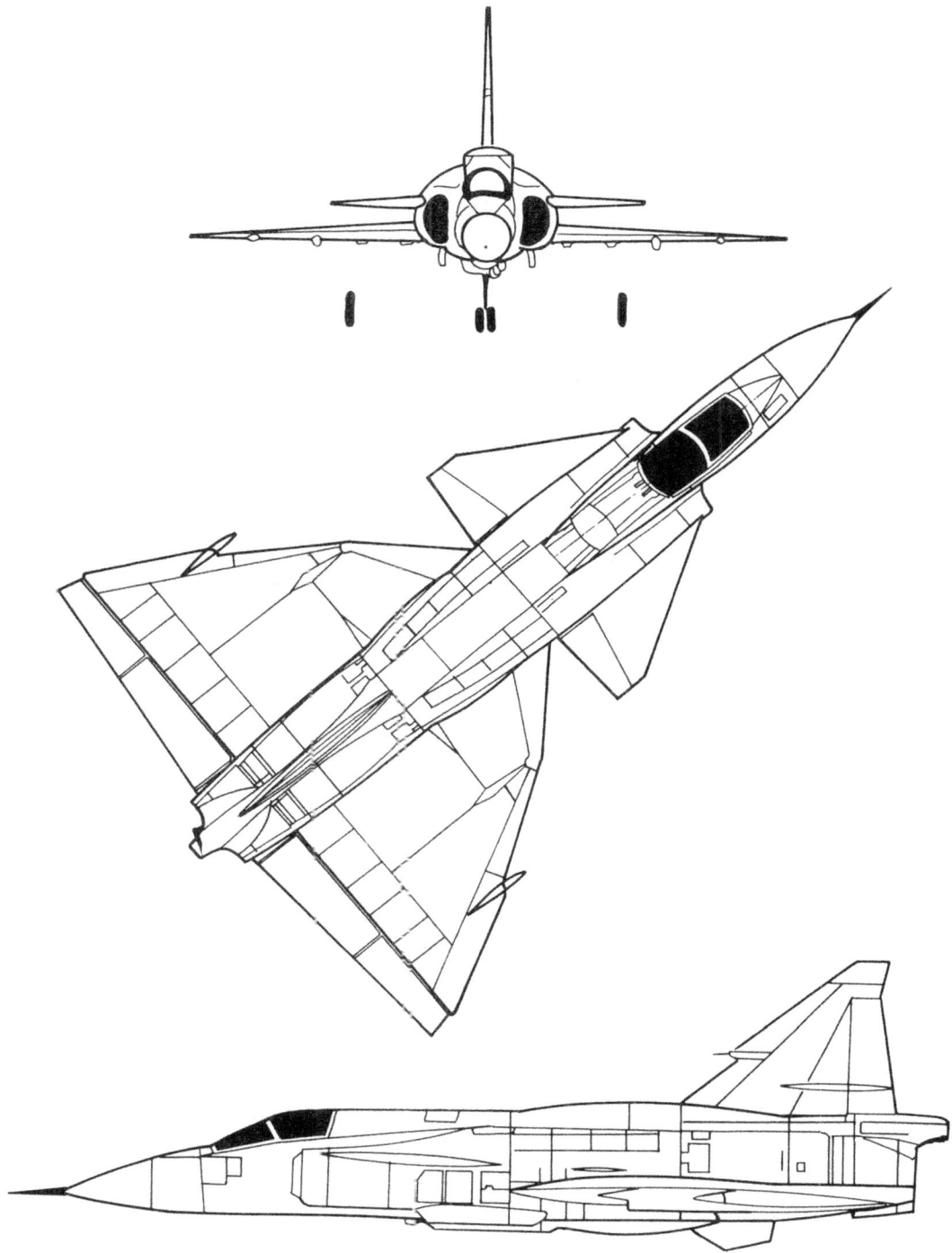

Fig. 3-2. Three-view drawing of the Saab Viggen canard fighter.

Fig. 3-3. The Saab Viggen taking off. Note downward deflected canard elevator.

7

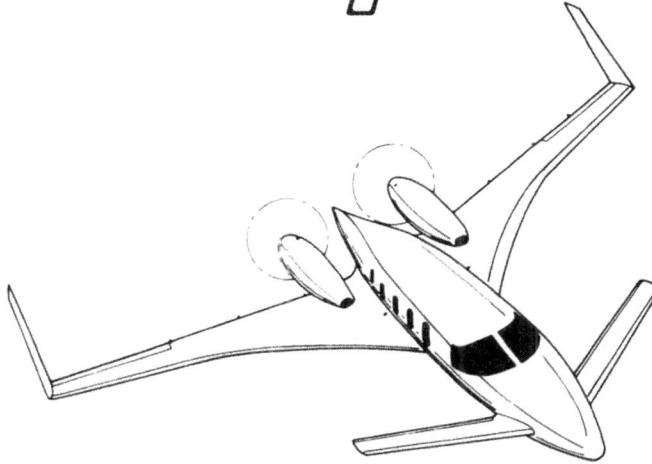

Light Civil Canards

LOCKSPEISER LDA-01 -
GREAT BRITAIN 1971

This very interesting airplane was the work of David Lockspeiser, formerly a British Aircraft Corporation pilot. It was specifically designed for bush operations, optimized from conception for primitive conditions. It is, in effect, an aerial utility van

Because of an easily accessible cargo area, and a large center of gravity range the LDA (Land Development Aircraft) is suitable for a wide variety of operations, such as freight, passengers, ambulance, supply dropping, crop spraying, fire fighting, patrol work and aerial surveying. It also has military applications. It could be fitted with floats, skis or wheels. The canard configuration was chosen because it offered a wide CG range, and freedom from stalling. The broad space between fore and aft planes also permitted ready loading and unloading, at a convenient height above the ground.

Aerodynamically, the LDA is unique. All flight controls are on the aft wing. Pitch and roll con-

trols are slotted flaps; the two inboard units serve as flaps, the outboard as ailerons. All combine in unison to operate as elevators. Since the wing cannot be stalled, these surfaces are always effective.

The foreplane is fitted with a slotted flap, in-flight adjustable as a pitch trim function, for various CG locations, and for trim when the aft wing inboard flaps are lowered for low speed landings. The vertical surfaces were aft wing mounted, with rudders at the rear for directional control.

Serious thought and attention was paid to the specification of construction, excellent flying characteristics and easy maintenance. It could be manufactured in those countries that desired to develop elementary manufacturing skills, and create employment, while at the same time, produce an aircraft designed for use in those same countries.

The wings are composed of three identical panels, of constant chord and wing section (NACA 23012), and constructed of light alloy stressed skin, riveted with pop-rivets and with parallel

Fig. 3-4. The Lockspeiser LDA-01 was built in Great Britain in 1971. It was designed as a light transport/cargo plane for developing countries.

main and rear spars. Each panel has four strongpoints in its center, that serve as attachment points to the fuselage when acting as foreplane, or as finpost attachments and lift strut attachment points when the panels are positioned as left and right aft wing panels. Flaps on both sets of wings are slotted and of sheet metal construction.

The fuselage is a welded, simple box-like structure utilizing ¾" square sectioned steel tubing, covered with a removable fabric bag. The pilot's compartment is paneled and faired below and ahead of the wind screen, with removeable fiberglass fairings that permit ready access to instruments and controls. A fiberglass fairing of clean design was also fitted to the aft 160 horsepower Lycoming pusher engine on the modified second version, as were Cessna-like, vortex controlling, wing tip fairings.

The wing mounted vertical surfaces are of welded steel tubing, fabric covered and wire braced. Balanced rudders extend above and below the wing trailing edges.

The landing gear on the first model was an unusual four wheeled affair with automotive hydraulic brakes on the rear wheels, and Ackerman steering on the front pair. The rear landing gear

struts were swept forward so that their internal structure would not reduce the useable space in the fuselage. The second edition utilized tricycle gear with a steerable, oleo-sprung nose wheel, and Cessna-like steel main landing gear stuts arranged vertically instead of swept forward, and with larger wheels.

The foreplane's angle of incidence is ground adjustable and mounted at 2½° greater incidence than the aft wing. It is fitted with leading edge stall strips to insure it stalls before the aft wing, due to its low aspect ratio compared to the rear plane.

Wind tunnel tests showed that the tip vortices from the foreplane disturbed the center semi-span of the mainplane at about 15° angle of attack. The fins and small wing fences were located to control this disturbance and insure safe, low-speed handling from the unstallable main wing.

Flight testing of the original version was extensive and led to modifications that will be discussed below.

Taxi runs up to 46 mph with foreplane trim-flap settings varying from 0° to 25°, showed this to be a powerful control. It significantly affected nosewheel lift-off speeds, for a given CG location, in the 40 to 52 mph range.

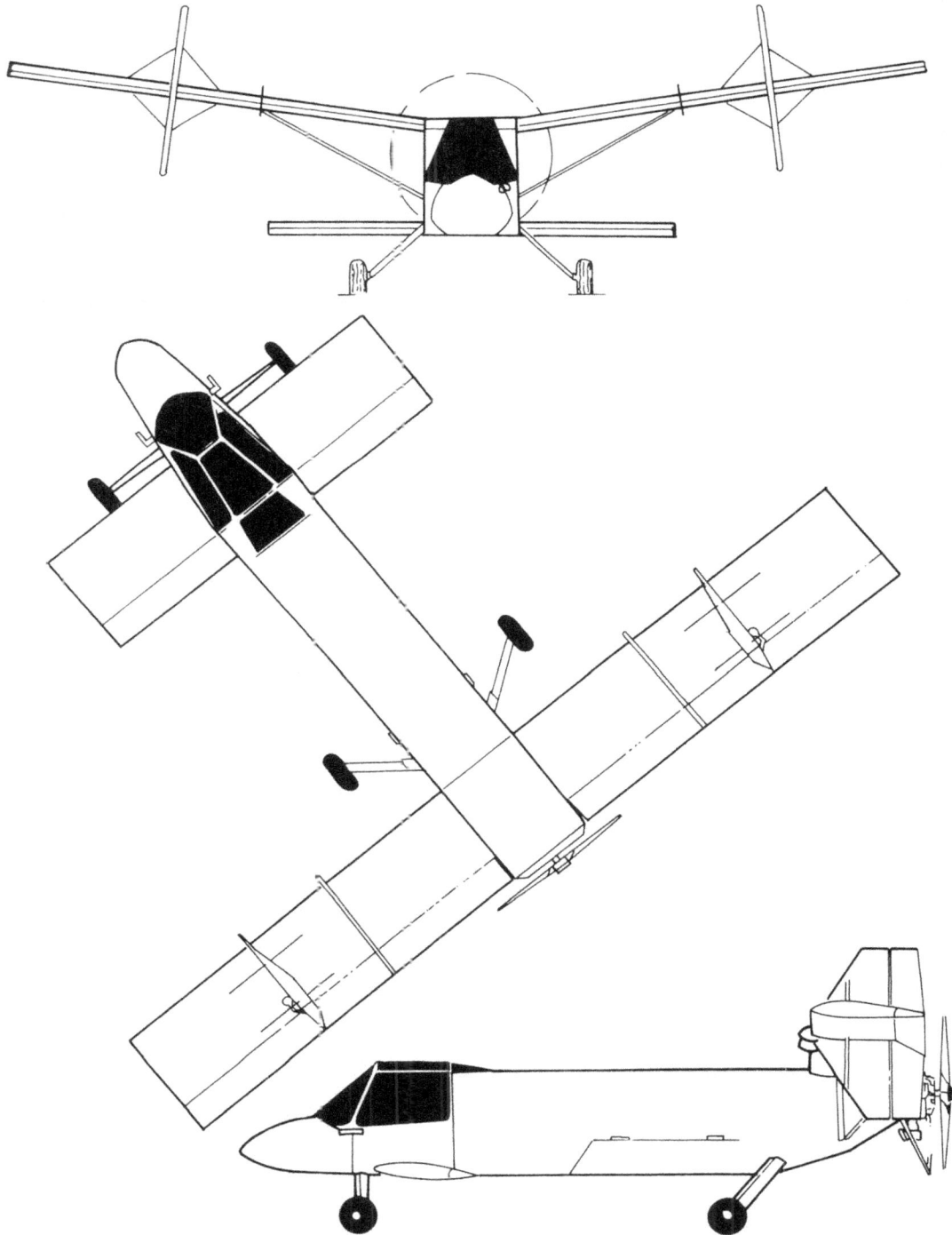

Fig. 3-5. Three view drawing of the Lockspeir LDA-01.

Takeoffs were normal with no tendency to swing, and the rudders became effective before nosewheel lift-off. Longitudinal control was positive and the degree of rotation and its rate were quite easily controlled. In flight, the aircraft was very stable, longitudinally. Control was light and positive, and it could be trimmed to fly hands-off. Mainplane flap extension gave a nosedown change of trim which was easily controlled by deflection of the foreplane flap, although simultaneous deployment of both fore and aft wing flaps would have been much safer, particularly when landing in adverse weather conditions.

The original configuration had lateral handling difficulties in coordinating turn entry and return to straight flight. On initiating the turn, adverse aileron yaw (no aileron differential) and roll rate caused adverse sideslip. Due to the large (7°) dihedral effect, the rolling moment due to sideslip opposed the applied aileron, neutralizing the roll rate.

The rudder was effective in suppressing sideslip, but light pedal forces made turn coordination difficult. An additional central fin, above the aft wing was tried, but it increased the dihedral effect as well as adding directional stability.

Handling modifications introduced in the second version included a dihedral reduction to 3°, and a six inch extension on each rudder to improve directional stability and control. Lateral control was greatly improved, although adverse yaw continued to give skid and slip on turn entry and recovery, but this was easily overcome by rudder action. Differential aileron action was not introduced. Rate of roll was proportional to the degree of aileron application.

Laterally, the aircraft is neutrally stable, although the deep fuselage necessitates a forceful effort to yaw the airplane into the wind during crosswind takeoffs. Spirally the aircraft, like most light airplanes, is mildly unstable.

The stall is gentle. The nose drops as the foreplane stalls, but with the aft plane still lifting. The front wing unstalls almost immediately, and the aircraft oscillates gently in pitch, stick hard back, while the aircraft maintains altitude at full power.

Cockpit visibility for the single pilot is excellent- and with the aft engine, vibration and noise proved low.

Lockspeiser displayed and flew his LDA-OI at the 1975 Paris Air Show. Considerable interest was displayed in his 70% of full scale prototype,

Fig. 3-6. The major components of the Lockspeir LDA-01 reveal its basic simplicity.

and he could have sold 100 of the full scale version had that quantity been available.

Rebuffed on financing by the British Government and London bankers, the LDA-01 was offered as a production package to countries with small industrial bases; along with setting up jigs and tooling for factory production. Judging from the very limited coverage of the LDA-01 in the Aviation Press since 1975, the prospect is either dormant or dead. This is unfortunate since this design offered so much. A larger turbo prop powered unit would seem very practical and useful as an aerial truck.

The projected full scale production aircraft was to have had a wing span of 39 feet - 6 inches, foreplane span 17 feet - 6 inches, and a length 31 feet - 10 inches. Its gross weight was to be 3,970 pounds and a disposable load of 1,985 pounds (or 100% of the aircraft's empty weight) was anticipated. Cargo space is a usable 250 cubic feet. The powerplant was to have been an air cooled unit in the 200 horsepower range, with a constant speed propeller.

LUSCOMBE "RATTLER -
GREAT BRITAIN 1983

The Rattler is being marketed as a very lightweight canard ground attack aircraft by Luscombe Aircraft of Kent, England; which was founded by Pat Luscombe, an ex-Royal Navy pilot.

It is designed to carry one 7.62 mm Chain Gun, mounted externally with the 2,000 round magazine located behind the pilot to avoid a CG change as the rounds are fired. Alternately, it can be fitted with rockets or a land mine dispenser.

The philosophy is that of a simple, inexpensive aircraft that can be crated to a forward area, and rapidly assembled by two men in 15 minutes. It would then be operated from grass strips or roads on close support missions. It could also be equipped as a pilotless drone for reconaissance of heavily defended targets.

The fuselage has a welded aluminum tube main frame. The wing is composed of a box section main spar, and an I-section rear spar with a polyurethane core. Fuselage and wing skins are of non-stressed fiberglass. Fuel tanks are in the wing center section and the airplane's belly.

The landing gear is tricycle, the foreplane is all moving, and the wing is slightly swept and tapered. Vertical surfaces are mounted largely below the wing trailing edge and conventional ailerons are employed.

Disassembly consists of removing the fins, unplugging the canard and removing the wing in two parts. Push pull control rods ease dismantling and assembly.

The empty weight is 450 pounds, with a payload of 650 pounds—or 45% more than the empty weight. The aircraft is stressed to +6/-3Gs.

The aircraft will not stall, but mushes downward at 31 mph. Takeoff requires 200 feet and 48 mph. The climb rate is 800 feet per minute. The airplane's maximum speed is 115 mph.

For short field landings, the rudders turn inward simultaneously to act as air brakes.

The engine is a Westlake 65/80 - 118-02, two cylinder 880cc displacement, producing 80 horsepower at 3,400 rpm and weighing 135 pounds - or 1.68 pounds per horsepower.

The dimensions are: wing span 36 feet, and length 14 feet - 6 inches. Total wing area is 160.27 square feet, of which the foreplane is 13.27 square feet, or 8.2% of the total area.

Luscombe forsees a potential middle east market of 400 aircraft.

PAT I PUGMOBILE -
UNITED STATES 1981

Howard "Pug" Piper who fathered the PAT-I project (Piper Advanced Technology) was the son of William T. Piper, founder of the widely known Piper Aircraft Co. After leaving Piper, he became an aviation consultant, but continued his interest in plastic and composite construction, dating from his work on the Piper Papoose plastic, two-place trainer of the early 60's.

After discussions with Burt Rutan, he became interested in the canard concept. Ultimately, he engaged George Mead, an associate of Burt Rutan's, and a talented airplane designer, to design and construct the airplane now known as the PAT-1 "Pugmobile."

This aircraft embodies a number of advances, including canard configuration, and an all-

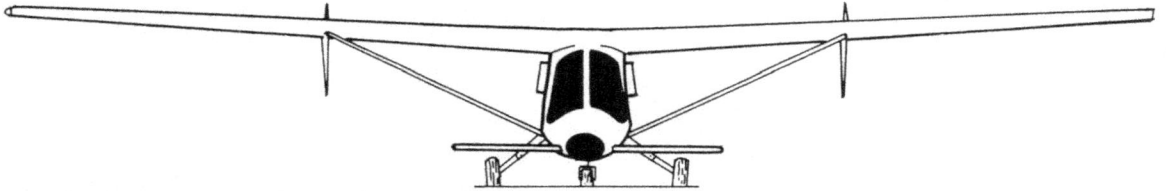

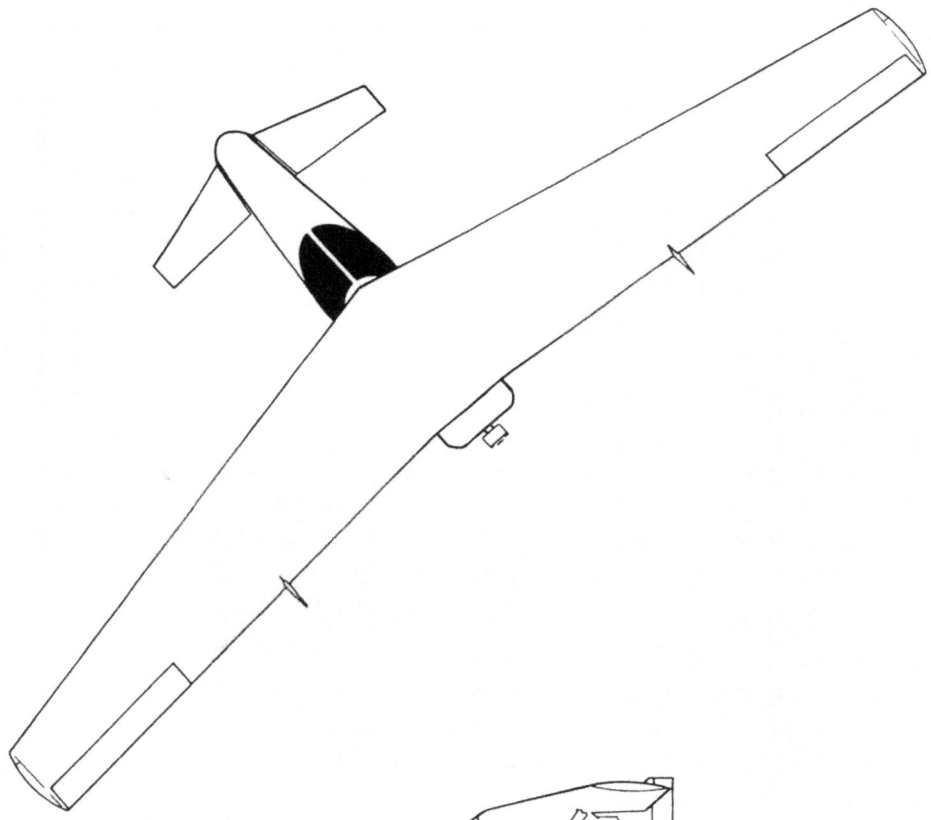

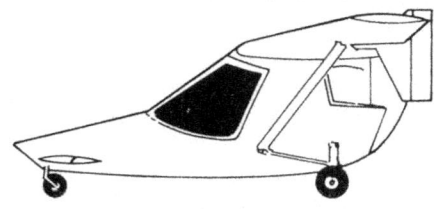

Fig. 3-7. The Luscombe Rattler was built in Great Britain in 1983. Its developer saw its role as a very light weight ground attack aircraft.

Fig. 3-8. The PAT-1 Pugmobile was built and flown in the United States in 1981. It was conceived as a possible production lightplane. Bud Davisson photo.

composite structure. It was designed from the start as a production airplane, not a homebuilt, and hence represents the first endeavor to apply the advanced technology developed for the amateur builder to general aviation aircraft. Mead's design calculations were verified on Burt Rutan's computer.

The aircraft itself closely resembles a conventional low wing aircraft. The foreplane, roughly one third of the total wing area in size, and equipped with a full span slotted elevator, sits just below the thrust line, immediately behind the tractor engine. The aft low wing, tapered and slightly swept, is below and just under the cabin, with its leading edge in line with the pilot. Judging from the main wheel position, the CG must be close behind the canard, which implies it carries a heavier wing loading than the main wing.

The fin and rudder are carried on the aft fuselage in the "normal position." The landing gear is tricycle, similar to that of the Gruman "American." The nose wheel is castered, steering on the ground being done by differential braking.

The engine is beautifully cowled, with a low set cooling air entry and upper air exits that are very much a Rutan configuration. Altogether, the impression is of a sleek, functional aircraft.

Design gross weight is 2,000 pounds with a useful load of 1,000 pounds. Dimensionally, it is comparable to a Cessna 172 and utilizes the same 150 horsepower engine and fixed-pitch propeller.

The fuel system consists of two 25 gallon tanks just behind the firewall. Fuel draw from the latter to the carburetor is via gravity, with engine-driven fuel-pump transfer from the wing tanks to the header. The wing tanks empty before the header and the CG moves forward as fuel is consumed. Should the engine fuel pump fail the header tank would provide an hour to land. The system requires no switching of tanks.

The airplane is very stable, with a cruise speed higher than the 172—about that of planes with 20 more horsepower, and very fuel efficient. There is no real stall. At extreme forward CG the pilot might run out of backstick. Most likely is a canard stall with a gentle nose drop. Approaches have been made at 75 mph, and stopping distances are less than 1,500 feet. The aircraft is very clean aerodynamically so that landings must be made at the correct speed; any excess means the aircraft will float over the runway before touching down.

Some form of drag device or flaps would be desirable to permit the slow, steep approach that, for example, a Cessna 172 can provide with its big barn-door slotted flaps. Since lift is shared by both force and aft wings, some form of slap would be required on both wings. On the rear wing only, deployed flaps would cause a dangerous and uncontrollable nose down pitch.

Composite construction, besides providing a very smooth external surface, means a great reduction in the many small components that go into the modern metal airplane. Instead, a relatively few big pieces go into the composite airplane. The technology for mass production is already available; fiberglass boats of all kinds, Corvette bodies, bathtubs, and lavatories are all made similarly.

Despite the promise of future greatness that the PAT-1 projects, its future is now very uncertain. "Pug" Piper died suddenly before its first flight and George Mead and two passengers were killed when the aircraft crashed. Hopefully the PAT-1 concept did not die, as well.

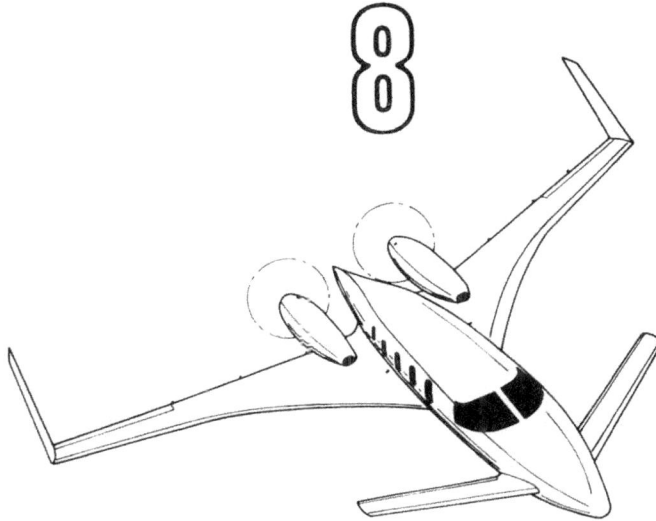

The Rutan
Amateur-Built Canards

Burt Rutan, in the estimation of the author, ranks along side the Wright brothers for his contribution to aircraft design as well as for his dedication to the "canard" concept.

Rutan is a graduate aerospace engineer. He worked first for the USAF developing stall/spin test procedures for advanced jet fighter aircraft. He also spent some time working for Jim Bede as flight test director on the financially ill-fated BD-5, and in his spare time, assisted NASA in stall/spin research. With this background, one can appreciate his concentration on the canard configuration, in all its variations, because of its inherent stall/spin resistance.

He set up the Rutan Aircraft Factory (RAF) in Mojave, California to develop canards, and sell plans for them to the amateur-built aircraft fraternity. In this, he has been extremely successful. More of his designs are now flying, world-wide, than any other brand of homebuilt aircraft.

In 1982, Rutan went into partnership with Herbert Iverson in an exciting new venture called "Scaled Composite Via Advanced Link to Effi-cient Development" or "SCALED" for short. The first effort in their new venture was a scaled down version of Fairchild's "New Generation Trainer" (NGT) built to Fairchild specifications. This NGT is Fairchild's entry into the USAF competition as a replacement for the current fleet of T-37 trainers. It was the successful competitor. Another "SCALED" development is the Microlight, a two place canard developed for Lotus.

Three of Rutan's designs have won the EAA's outstanding new design awards, the Vari-Viggen (1974), VariEze (1975) and Quickie (1978), a singular tribute. The following will deal with seven of Rutan's beautiful designs.

VARI-VIGGEN, MODEL 27

This is a two-place canard pusher whose design commenced in 1961. It was built in 1968, after wind tunnel testing, car-top experiments and finally, a larger radio controlled model which was flown extensively.

As the name implies this aircraft owes some of its conception to the Saab-Scania "Viggen" re-

Fig. 3-9. The original VariViggen of 1974 (foreground) is flanked by Rutan's Long-EZ, both pioneers in the modern movement towards canard aircraft.

viewed earlier in this book. The Vari-Viggen has a delta main wing, recessed at the center to clear the pusher propeller. Twin fins and rudders projecting behind the wing trailing edge provide directional stability and control. Ailerons are located outboard of the fins. The foreplane is rectangular in shape, of low aspect ratio and carries the pitch control, a slotted flap, on its trailing edge.

The tricycle landing gear is electrically retractable. The main gear incorporates knee-action trailing beams with rubber shock discs, reminiscent of an Ercoupe undercarriage, and retracts inward into walls in the wing underside. The steerable nose wheel retracts forward into the fiber-

glass nose cone.

The engine is a Lycoming O-320 of 150 horsepower and turns a metal, two bladed 75 inches diameter, pusher propeller. The plane seats two in tandem, the pilot up front and the passenger behind—right on the aircraft center of gravity, thus preventing any CG shift when flown solo. Structurally, the airplane has an all wood stressed skin except for the wing panels outboard of the fins, which are aluminum. Ailerons are mass balanced for flutter prevention.

Rutan's major design objectives were stability, safety and controllability at low airspeeds, plus the absence of stall/spin. In this he succeeded admirably—the aircraft will fly with full back-stick at 50 mph with full roll control. Turns require little coordination of rudders and ailerons; spiral stability is neutral visibility over the canard is excellent and the airplane cannot be spun.

There are no high lift devices to reduce speed or steepen the glide path, but the plane can be side slipped, power-off, to reduce both speed and altitude. The landing is normally nose high.

While primary pitch control is via the forward wing's slotted flaps, the ailerons are equipped with a slow moving, electrically driven reflex action operated by a thumb button on the control stick. Full differential action as ailerons is retained. Reflexed downward, a cruise speed gain of almost seven mph is achieved. Upward reflexing permits lower rotation speed for takeoff and lower landing speed than downward reflex.

The high position of the canard, relative to the aft wing, permits the wake centerline from the foreplane to flow above the aft plane, contributing to the latter's lift.

The Viggen is designed specially for home construction with few tools. Removal of the outer wing panels and nose gear retraction, converts the aircraft to an eight foot wide trailer for backyard or garage storage. A maximum of flat surfaces, with compound curves limited to the fiberglass nose cone, engine cowling tank covers, and canopy, makes construction relatively easy. Four

Fig. 3-10. The VariViggen design (1975) was inspired by the Saab Viggen canard fighterjet. The Sp Model 32 pictured here, featured extended, tapered wing tips, and Rutan's first use of Witcomb winglets. These modifications increased speed 5 - 7 mph, while rate of climb went up 25%.

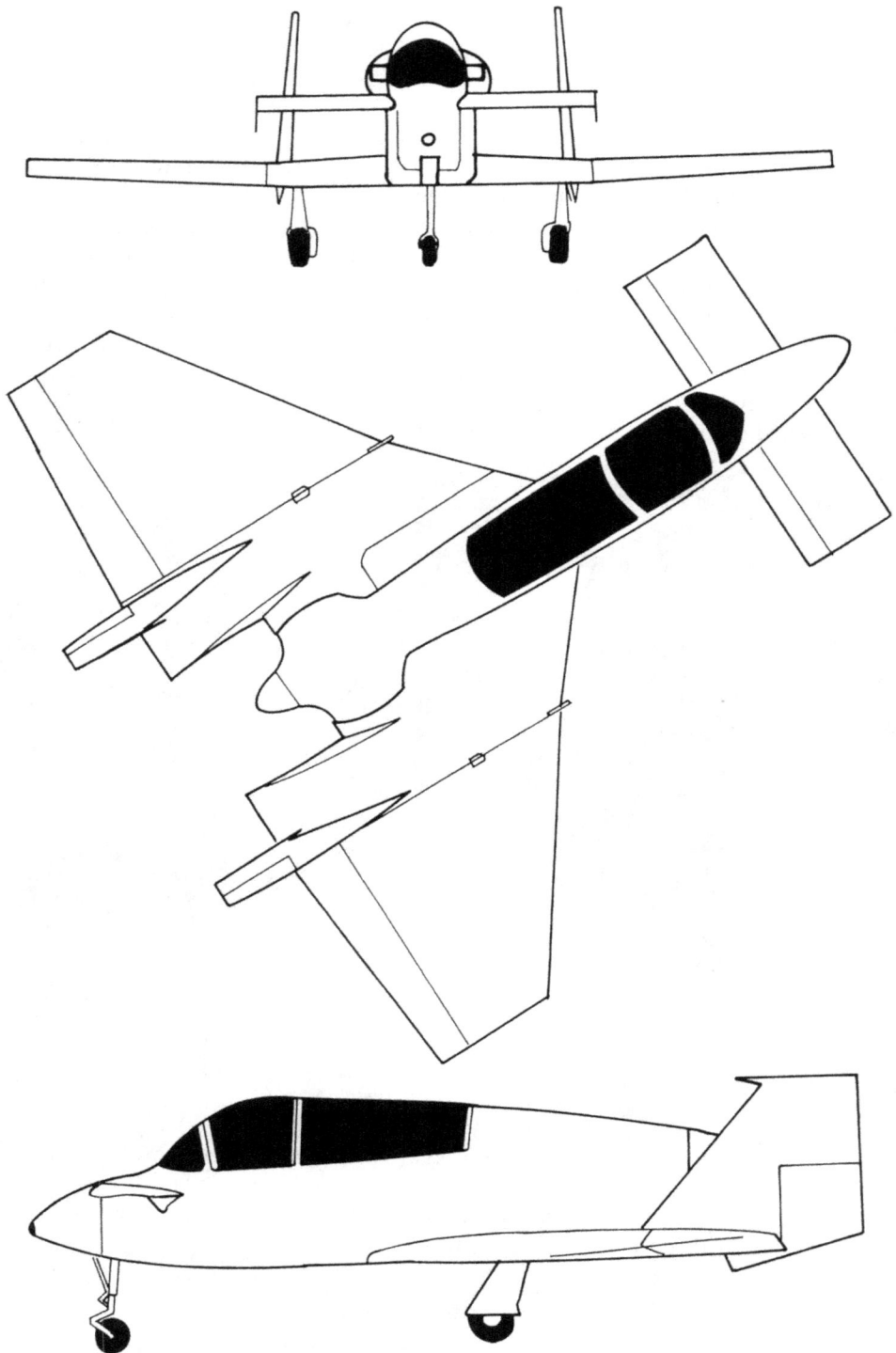

Fig. 3-11. Three view drawing of Vari-Viggen

years of spare time construction is required to construct a Vari-Viggen.

The Vari-Viggen SP Model No. 32 followed, which incorporated new outer wing panels of Wortmann FX 60-126 wing section built of fiberglass over urethane foam construction. These increased the wing area to 125 square feet and span to 23.7 feet. The rate-of-climb increased by 25% while cruise went up five to seven mph.

VariEze, Model No. 33

In his search for a lower cost homebuilt aircraft of lower horsepower, but with excellent performance, Rutan in 1974 started to design and build the "Mini-Viggen." This was a high wing, low canard, all metal two place tandem aircraft powered by the then available Franklin 60 horsepower aircraft engine.

The wings had a moderate sweep back and dihedral, were of constant chord, with wing tip rudders and fins projecting downward. The canard was rectangular and the tricycle landing gear was fixed, but intended to be well streamlined with wheel fairings. Ailerons at the aft wing provided roll control and the canard flap elevators provided pitch control.

This partly built prototype was abandoned after finding the metal structure too complex and heavy. Also, model testing indicated spiral instability at approach speed.

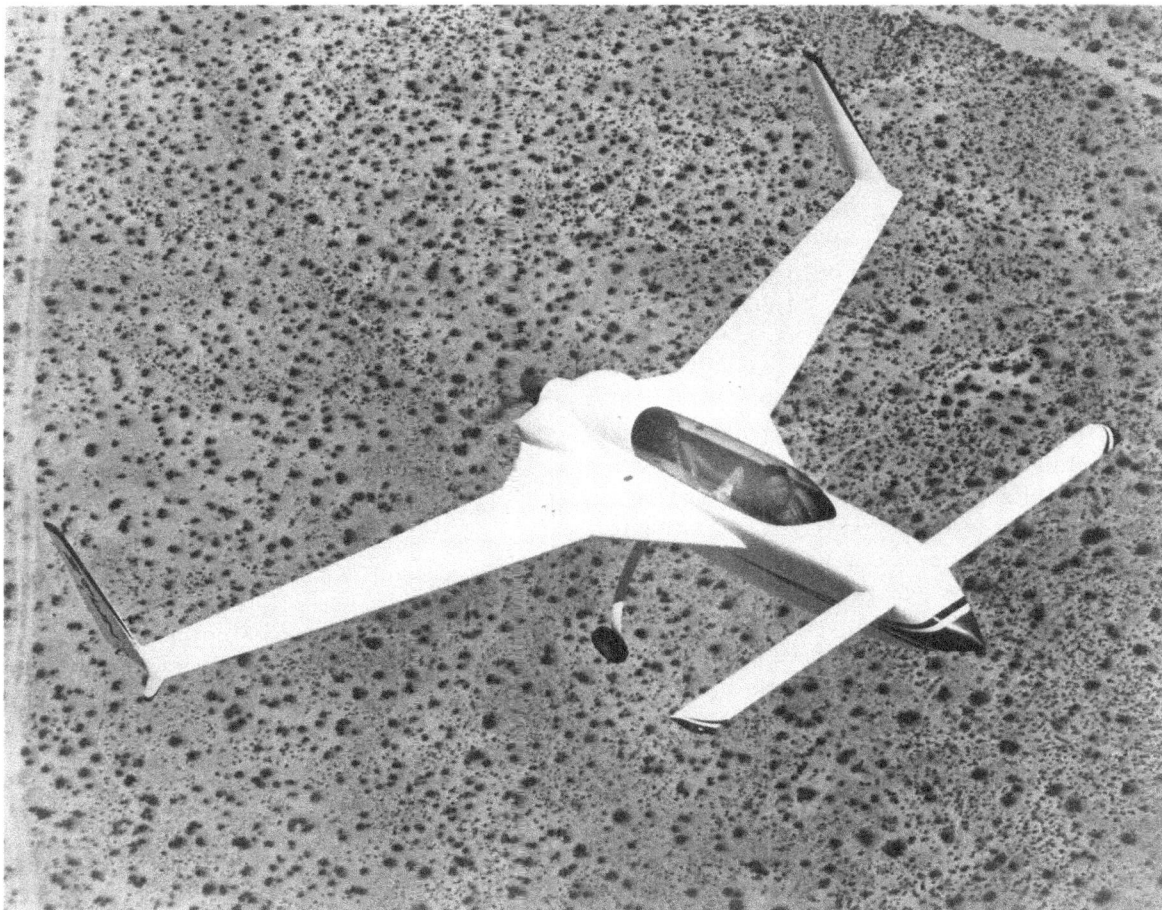

Fig. 3-12. The Rutan Model 33 VariEze of 1975 is what really set the world on its ear. Not only did it popularize the canard concept among amateur builders, but it introduced hem to composite construction techniques as well.

The next project was the VariEze Model No. 32, designed to use a Volkswagen engine of 62 horsepower, weighing 139 pounds.

About this time, Rutan became very interested in European sailplane structures of smooth contoured, efficient, glass composite materials. This was considered to be the most significant advance in light plane construction and aerodynamics. The light weight, smooth structures, improved corrosion resistance, lengthened fatigue life, had much fewer parts, and were ideal for homebuilt construction.

Rutan then realized that he needed to develop methods for building composite structures using only the basic lay-up, *not* the complex European mold lay-ups. He would have to use materials proven through years of aircraft experience.

Many test samples were built for strength, temperature cycling, stiffness, creep, surface durability and fuel compatibility. Larger samples were tested on material testing machines at Edwards Air Force Base.

Rutan decided to use this all-composite-glass-foam sandwich structure in the design of the first VariEze prototype. It would be used as a research airplane for both structures and aerodynamics. One of the major virtues of his structure is that the high strength material is placed as far from the neutral axis as possible for strength with light weight, plus the very important objective of quick construction, so vital for the development work Rutan was embarking on. The result was a structure of excellent surface durability, a 12-g ultimate load factor, and competitive weight.

The performance and efficiency of this VW-powered airplane was excellent. Top speed was over 180 mph. At high cruise speed it achieved 40 mpg, while at 95 mph it got over 60 mpg. However, the poor reliability of the VW version being used, and its high maintenance requirements, rendered it unsuitable for the VariEze.

In addition to the engine problem, there were other difficulties. The stall and landing speeds were too high, roll control (combined pitch and roll control from the canard elevons) was poor at low speeds, and the nose-wheel retracting system was unsatisfactory.

In the design of this prototype, Rutan had certain aerodynamic qualities as objectives. First, was to provide natural, passive angle of attack limiting to make the aircraft departure and spin proof. The then new Whitcomb winglets would reduce induced drag for better climb and cruise efficiency. The use of canard elevons for both pitch and roll control resulted in a simple control system. The lower wetted area accommodation for two occupants in tandem was better than conventional designs.

Fuel tanks in the wing root strakes plus a passenger on the CG resulted in little CG shift. The rudders incorporated in the winglets, operating outward only, provided effective directional control

Changing the canard wing section from the GAW-I airfoil to the new University of Glasgow developed airfoil section GU 25-5 (11)8, corrected the stall speed problem, lowering it by eight mph. Roll control with elevons was greatly improved at low speeds. The nose gear problem was overcome by deciding that the pilot should raise the nose from the kneeling position, now typical of the VariEze, before boarding the airplane. The nose gear was simplified to include up and down loads.

The engine problem was cured by selecting the 173 pound Continental A-75, but this required a scaling up of the whole airplane from 59 square feet to 67 square feet of wing area—calling for a completely new prototype. This was just as well, since the original had been built without tooling. For a homebuilt however, tooling had to be provided for certain essential components. Design changes were also made to permit use of 0-200, 0-90, 0-85, A-75 and A-65 Continental engines, plus adaption to the 1700cc VW conversions.

Other changes being made to the homebuilt prototype included a shallower fuselage and taller canopy, plus a roll-over structure behind the pilot. A longer nose, more rear seat room, a larger instrument panel for full IFR instrumentation were incorporated. Revised side stick controls, increased fuel capacity, increased canard aspect ratio, larger main gear tires, better brakes, and simplification of both fuel system and structure to reduce the quantity of parts, were also included.

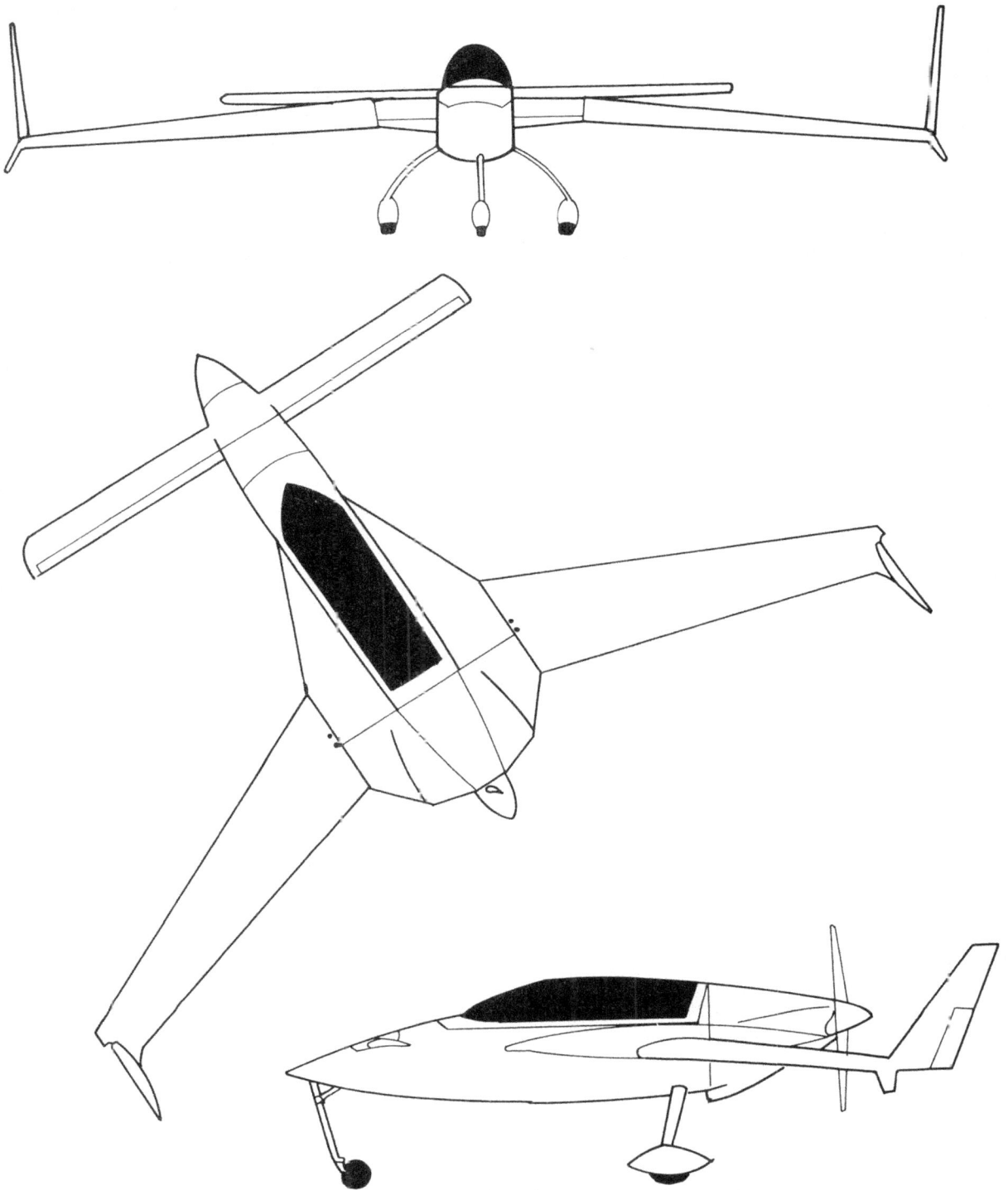

Fig. 3-13. Three view drawing of VariEze.

The one piece fixed main landing gear was retained, but well faired. The one piece design eliminates the high leverage loads on the attachment structure that would be inherent in a two piece unit. The reclining pilot and passenger position reduce both frontal area and fatigue in long flights.

Aerodynamically, a moderately high wing loading raised the speed of best lift-to-drag ratio, while providing a better ride in turbulence. Winglets reduced drag 15%; burying the engine in the wing roots and avoidance of any surface protuberances, plus nose wheel retraction, reduced parasitic drag. The landing speed of 74 mph was still high.

The aircraft has mild positive spiral stability, and speed stability with no pitch, roll, or yaw trim changes due to power changes. It can be flown hands-off for extended periods.

It can be disassembled into four parts for easy road haulage and storage at home. The aircraft was not designed for aerobatics.

An 85 hour test program on the homebuilt prototype was completed during a ten week period. It included flying qualities, performance determination, dive and spin tests, and environmental qualifications. It would not stall/spin, and flying qualities were satisfactory. A range of 700 miles with useful load and two people was achieved.

After the introduction of the VariEze to the amateur-built aircraft market, the response was enthusiastic and as the "hatchlings" started to emerge from garages and basements, so did problems. The most serious was that a comparatively minor rigging error in rearwing incidence and twist, could overpower the roll control ability of the elevons, and called for full roll control plus rudder just to stay upright.

Aileron deflection of the elevons caused a downwash change which impacted on the aft wing in such a way as to oppose the elevon roll input. Additionally, when the elevon was deflected as an elevator at slow speed, the aileron action was poor. It was decided to use the elevons as elevators only, and to incorporate proper ailerons on the aft wing.

The ailerons were designed, and tested throughout the flight envelope, including stall/spin and flutter. Test reports indicated that the stall proof qualities were being realized on homebuilts. No departures from controlled flight occurred even in sideslips or full-aft-stick flight, power on or off.

However, some builders were experiencing divergent wing rock at the stall, or roll off at aft CG, but within CG limits, at low speeds. This was explored on the prototype. The correction was leading edge droop or cuffs located on the aft wing outboard leading edge which cured the aftwing tip stalling tendency, so characteristic of swept, tapered wings.

Another problem, was collapsing of the nose wheel when the down lock failed. Redesign of the nose wheel retraction system to incorporate a worm gear mechanism eliminated the problem and permitted a 30 mph increase in gear extension speed.

To permit steeper approaches without speed build-up, a belly drag-brake was devised that is evident in the photo of N4E2 just above the runway.

Long-EZ, Model 61

In 1978, Rutan surveyed the VariEze and its market position to determine if its long range prospects were suitable. After assessing the situation, he decided to redesign the aircraft for one major reason. Many builders were being forced to use the O-235 Lycoming Engine due to both scarcity and high price of the Continentals. This engine being heavier, even without starter and alternator, would render the VarEze tail heavy and overweight.

It was decided to design the new aircraft around the Lycoming O-225 with starter and alternator, including a complete electrical system with lighting. The O-200 Continental engine would also fit the new design. Design objectives were long range, good forward visibility, and lower approach and landing speeds more suitable for low-time pilots. The same stall-free features of the VariEze were mandatory. The aircraft was built in four months and made its first flight on June 12, 1979—but it did not fly well.

The configuration used VariEze outboard wing panels mounted on a wider centersection, and with greater sweepback to compensate for the heavier engine. No rudders were incorporated in

Fig.3-14. The Rutan Long-EZ appears identical to its predecessor, the VariEze, but it is actually a completely new, larger design. Field experience with VariEze builders led to the need for this new aircraft, which could take standard aircraft engines. It offered greater range and payload capabilities, as well as reduced landing speed.

the winglets, but a "rhino" rudder was placed on the nose, for control simplicity.

Directional stability was weak, dihedral effect too great; adverse yaw was high; roll rate was inadequate and the stall speed was too high. In addition, early airflow separation on the main wing caused instability in pitch at low speed.

Despite extensive modifications, some of which were successful, and much test flying, the stall speed remained high, landing attitude was poor, and the roll rate remained unsatisfactory. By August 1979, it was evident, that a new aft wing design would have to be developed and a new wing built.

First flown in October 1979, the new aft wing had less sweep, greater area, and a new airfoil, developed by Dr. Richard Eppler, a widely known German aerodynamacist. It also featured longer ailerons, a better winglet-to-wing tip fairing to eliminate airflow separation at the wing tips, and an outboard wing attachment that permitted incidence adjustment.

It performed very well in tests. Approach and landing speeds were lowered, full stall landings with good visibility were achieved, and the roll rate was excellent. However, weak directional instability led to removal of the "rhino" rudder, and installation of larger winglets, incorporating outward deflecting rudders, correcting the problem.

Complete flight tests in December 1979 proved highly successful. It has demonstrated resistance to departure from controlled flight in all kinds of stall entries, including slides. It is firmly stable even at full aft CG. While its wing area is 41% greater than the Vari-Eze with a 26% higher gross weight, its cruise speed at 75% power is 183 mph, only 14 miles slower than the earlier design.

The Long-EZ statistics are as follows: gross weight 1,425 pounds, empty weight 750 pounds, and useful load 675 pounds (or 90% of empty weight). Its fuel capacity is 52 gallons, range at gross and 75% power is 1,250 miles, and at 40% power is 1,970 miles. Landing speed solo is 59 mph, and 68 mph gross. Ceiling is 22,000 feet.

Like its predecessor, the Long-EZ has a high aspect ratio forward wing with slotted flap elevators, retractable nose wheel, well streamlined main gear, and a typical nose down posture at rest. The engine wing roots are beautifully faired, and the wing leading edge has a double crank near the fuselage, whereas the VariEze had only one.

Like the Vari, the Long incorporates separate pedals for directional control and braking. The nose wheel is free castering so ground maneuvering requires differential braking action. Another feature—the twin pedals—permit both rudders to be deflected outward, as a form of dive brake, for steeper descents. Light pedal pressure gives rudder action; heavier pressure actuates the brake— together for slowing down and differentially for steering on the runway.

The author flies a Piper Warrior and prefers to use the brake hand lever for slowing down. The toe brakes, even after many hours of flight, continue to feel awkward. The Rutan pedals seem more "natural."

Rutan's market appraisal was correct. The Long-EZ is the most popular homebuilt yet. Within two months of plan availability 500 sets were sold, and sales continue briskly.

On December 15, 1979 Dick Rutan, Burt's brother, broke the closed-course world distance record, for aircraft of the Long-EZ's weight, by flying 4,800.28 statute miles in 33 hours, 33 minutes and 41 seconds at an average speed of 145.7 mph. Fuel economy was 35 miles to the US gallon. Auxiliary fuel tanks gave a total capacity of 143.6 gallons of which, 3.75 gallons remained at the flight's end.

This record breaking flight was officially witnessed and certified by both the Federation Aeronautical International (FAI) and the National Aeronautical Association (NAA) in recognition of this great achievement in design skill and piloting by the Rutan brothers.

Defiant, Model 40

The Defiant first flew on June 30, 1978 after 7½ months in design and construction. In typical Rutan fashion, a minimum of paperwork or drawings was prepared for its construction.

The Defiant is powered by twin Lycoming O-320, 160 horsepower engines, located at the extremities of the fuselage—one in the nose

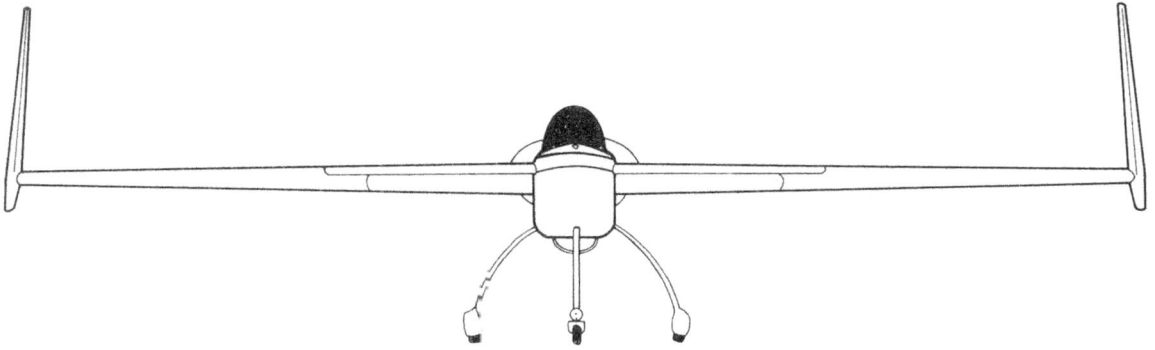

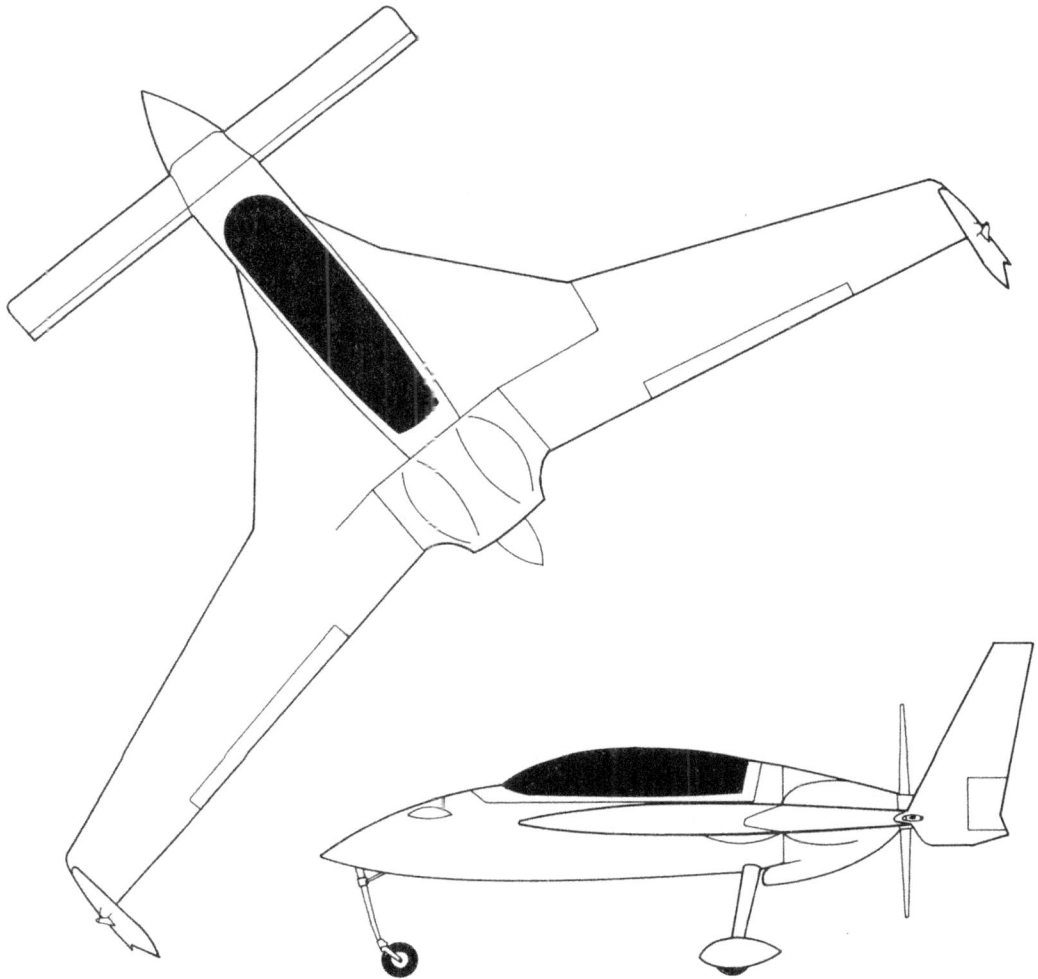

Fig. 3-15. Three view drawing of Long-EZ.

Fig. 3-16. The 1978 Rutan Model 40 Defiant, was Rutan's solution to a safe, efficient twin. Centerline thrust offers the pilot a procedureless transition to single engine flight, in the event of an engine failure. The aircraft cruises at 216 mph at 70% power, while stall occurs at 74 mph. The panel will take full IFR instrumentation.

pulling and one in the rear pushing. It is a four place aircraft.

The aircraft's statistics are as follows: Length 32 feet, span 29 feet, total wing area 127 square feet, wing loading 23 pounds per square foot, power loading nine pounds per horsepower, empty weight 1,525 pounds, gross weight 2,900 pounds and useful load 1,375 pounds or 90% of its empty weight. Fuel capacity is 1,590 gallons. Cruise at 70% power is 216 mph, and 195 mph at 35% power. Stall occurs at 74 mph, while landing speed is 80 mph.

Starting from the nose of the airplane, and working back, the front engine is positioned just ahead of the canard wing. The engine cooling air inlet is below the spinner surrounding the hub of the 69 inch diameter Kevlar covered, wooden fixed pitch propeller. An eight inch extension shaft permits better cowling lines.

Cooling air exits are in the top of the engine cowl, exiting into a low pressure area. The combination of ram air in and reduced pressure out, eliminates the need for cooling flaps, even during low airspeed, full power climb conditions.

The canard is a high aspect ratio unit, with slotted flap type elevators, and tips with swept back leading edges. The final version employs an original Rutan section.

Directional control is via a forward rudder whose virtue is a very simple control linkage. In the event of a nose-gear-up landing, it would be destroyed.

The nose wheel retracts into a streamlined fairing on the fuselage underside centerline. The gear is a Mooney nosewheel unit.

The fuselage is of rectangular cross section with well rounded corners. Considering that it contains two engines, space for an instrument panel and

116

four large people and baggage, there is simply no wasted space—such as in the tail cone of a conventional aircraft.

Entry to the cabin through the swing over canopy is awkward, requiring one to mount the aft wing stroke. The latest version has easier access via steps and hand holds.

The canopy, forming the major portion of the cabin roof, was originally round in cross-section, but will be flat sided with generous corner radii, allowing for improved spaciousness.

The main landing gear is typically bow-legged, rugged and swept forward somewhat.

The aft wing has the distinctive strakes running well forward, almost to the canard. They hold the fuel, and are arranged so that very little CG change occurs with fuel burnoff.

The wings are of Eppler section, well swept back, with winglets at their tips for directional stability, but without rudders.

The ailerons are uniquely Rutan. They are set into the wing trailing edge on either side of the fuselage. While they do not have the leverage of wing tip ailerons, their location has major advantages. They minimize wing twist and adverse yaw, but do not induce tip stalling. Their location also permits a simple in-fuselage control linkage. The roll rate is not that of a Pitts Special but, who needs that on an airplane designed for moving people in comfort, at high speeds.

The aft engine is pusher mounted, with cooling air intake offset to avoid ingesting front engine exhaust fumes. A generous cooling air outlet is provided below the aft spinner. The exhaust of the rear engine exists from the rear open end of the cooling air intake fairing. The Kevlar covered wooden pusher propeller is of fixed pitch, two inches greater in pitch than the front prop. Like the forward engine, no cowl flaps are employed. The aircraft has no high lift devices on the aft wing to permit slow, steep approaches.

Apart from the complete absence of the aft wing stall which permits full aileron roll control, the centerline thrust, fixed pitch propellers and absence of wing flaps and cooling flaps, render the aircraft much safer than a conventional light twin engine airplane. When one engine in the Defiant quits, the pilot literally does nothing but continues to fly the airplane.

The pilot of the wing-mounted-engine twin becomes a very busy man if he wants to live. He has to identify the failed engine and feather its prop, obtain maximum power from the active power unit, open cowl flaps, monitor control speed, retract his landing gear and flaps if going around or on takeoff, and above all fly the airplane, compensating for offset thrust with rudder and aileron.

The Defiant has two separate and independent power systems. Each has its own fuel tank with a cross feed system. Each also has its own alternator, regulator, battery, and busbar. The electrical load of radios, lights, etc. are shared, but an electrical cross feed permits one unit to operate all electronics and charge both batteries. However, since the failed engine propeller windmills at around 1,000 rpm, neither the alternator or vacuum pump is lost.

There are no trim changes with power, no minimum control speed and no stall/spin, regardless of power setting or control position. The efficient use of space permits a lower gross weight, which provides good performance and climb on one engine. Steep turns with full backstick are routine.

Engine controls are on a central console. The flight controls are side arm wrist sticks on the right and left cockpit sidewalls, with convenient and comfortable arm rests. The Kevlar wooden propellers and fiberglass airframe are quiet, producing very little annoying out-of-sync noise.

At forward-limit CG, takeoffs require considerable speed for the canard to cause rotation. Similarly, on landing the nose wheel contacts the runway fairly soon after touchdown to the canard's heavier wing loading.

Landings require a long flat approach to avoid speed build up and resulting float. As a pilot who likes a steep, slow, fully flapped approach, the author feels this, plus the relatively fast touchdown, are the only points open to criticism. Some form of drag brake would seem desirable, but this is a very minor complaint in the light of this airplane's tremendous merit.

Rutan intended the Defiant to be a "production" airplane rather than a homebuilt. As such, he exhibited it first at the National Business Aircraft Association (NBAA) meeting in 1979. He received many offers of financial assistance from various individuals and groups, to obtain certification leading to volume production. Such an undertaking would have meant that Rutan would lose the freedom to do what he likes best, design; and these financial backers would have demanded alleged improvements, such as fully retractable landing gear, variable pitch propellers, etc. and the basic simplicity of Rutan's design would have been lost.

He did contemplate certification on his own, but this was ultimately ruled out. Rutan then contracted with Fred Keller of Anchorage, Alaska, to develop plans, build a revised prototype and prepare bills of material for the purpose of offering the Defiant as a homebuilt project. This answers the pleas of many EAA'ers who have ached to build this airplane.

The prototype will incorporate some modifications, as follows:

- Wing span increase of 27 inches, and canard span increase of 24 inches
- Relocation of winglets aft by several inches, while canting them inward 8°
- Reposition of aileron hinge points for a better roll rate
- Wider opening canopy supported by gas springs
- Better step location for easier entry and exit, particularly for skirted ladies.

Rutan's Defiant is undoubtedly the most important single break through in general aviation for many decades. It well deserves the name "Defiant."

There is also a "Baby Defiant," called the "Gemini" with side by side seating and twin pull push type IV VW engines in a roughly LongEZ sized package, designed by Dave Ganzer. It had its configuration, canard geometry, airfoil sections, etc. run through Rutan's computer program— a very wise move.

Grizzly, Model No. 72

The VariEze lands at 74 mph; the Defiant at 80 mph. These are fairly high speeds and require long shallow approaches to land "on the numbers." This was recognized in the Long-EZ design,

and is reflected in its lower wing loading and landing speed of 68 mph.

Contrast these speeds with that of a Cessna 172 landing at 55 mph with full flaps, which also permit a steep approach without speed buildup. To the author, the approach and landing characteristics of Rutan's designs are the only area open for mild criticism, of a constructive nature, of course, the Grizzly reflects Rutan's awareness of this condition.

A canard or tandem wing airplane has two lifting surfaces. Application of high lift devices is thus a major problem. Not only must the added lift be distributed so as to permit fully controlled flight on a level keel, but also the "no departures," no "stall/spin" characteristics must be preserved. This all leads to both design and control linkage problems, not easily resolved.

On January 14, 1982 Rutan unveiled his model No. 72 "Grizzly," a hybrid which incorporates huge fowler flaps on both fore and aft wings. The airfoil and wing systems of this aircraft were the first application of a new aerodynamic design computer program developed to handle the very complex aerodynamic relationships of tandem wings and high lift devices, both in flight and in ground effect.

The Grizzly is intended as a proof-of-concept research vehicle, to investigate the STOL characteristics of tandem wing aircraft, new amphibious floats and a number of structural concepts. No plans exist for its production, or for marketing it as a homebuilt.

A description of the aircraft starts at the nose and ends at the tail. The power unit is a Lycoming L0360B 180 horsepower engine turning a constant speed, two blade Hartzell propeller with Q-tips. The engine cowling and spinner reflect the Defiant design—an oval cooling air intake below the spinner, with air outlets on the cowling top rear, exiting in an area of low pressure, particularly in the climb. No cowl flaps are needed.

The aircraft is a "taildragger." The main landing gear consists of two elephant-tusk-like legs, curved backward, each carrying a pair of low pressure tired wheels mounted on a hub at the bottom of the legs. This gear is obviously intended for heavy duty, and incorporates disk brakes.

Fig. 3-17. The Rutan Model 72 Grizzly is a proof-of-concept research aircraft with STOL capability. It was designed in 1980, built during 1981, and first flew in 1982. It is a four-place tractor with swept-forward canard and main wings. It features 55% chord Fowler flaps on each wing, plus an aft elevator.

The forward wing is of constant chord, and is swept forward about 10°. Huge Fowler flaps of roughly half the foreplane area cover most of the trailing edge. The remainder is composed of small

elevators between the inboard flap tracks and the fuselage side wall, operating in concert with the horizontal tail.

The flap tracks are of simple design. As the flaps

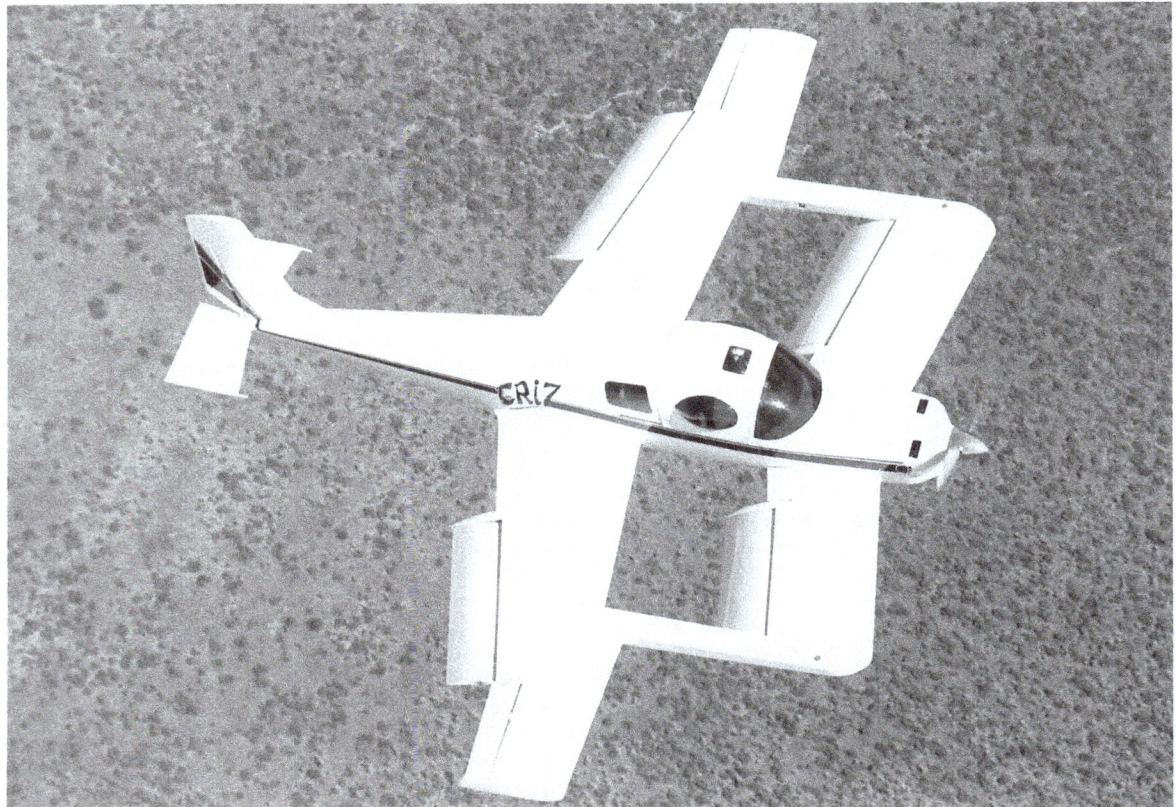

Fig. 3-18. The Rutan Grizzly with Fowler flaps fully deployed.

move back, their angle of attack is controlled by an extension of the inboard track in a streamlined support. The outboard end is not similarly supported; the rigidity of the foam/fiberglass structure permits this three point suspension.

Running backward from the foreplane tips are two rectangular units that merge into the aft wing, providing rigidity for both wings and serving as fuel tanks.

The cockpit seats four, and has a fixed windshield of beautiful proportions. The canopy slides back on tracks and has two clear, streamlined blisters, permitting downward vision.

The fuselage extends backward to support a large fin and rudder, and an all moving elevator that appears to have a thin inverted lifting airfoil cross section.

The aft wing, between the fore and aft booms, is also swept forward, but is tapered and carries large Fowler flaps of similar design to those on the foreplane.

Outboard of the booms, the wings are not swept forward, and their trailing edges contain slotted ailerons. Vertical dividers act to seal the inboard ends of the ailerons from the airflow around the outboard ends of the deflected Fowler flaps.

The aft end of the fuselage supports a tail fin with dorsal, as well as the rudder. The all-moving elevator has swept forward leading edges; is hinged in the middle and mass balanced for flutter prevention. The straight trailing edge carries an anti-servo tab for control feel and longitudinal trimming. A large steerable tail wheel is mounted on a projection of the vertical fin, located below the fuselage; the wheel holds the fuselage aft end well off the ground to reduce the nose-up angle of the fuselage center-line.

The Grizzly is all-composite, using fiberglass and carbon fiber for added strength. One wing is constructed of full core/fiberglass similar to the VariEze; the other is skinned with a fiberglass/foam sandwich type surface leaving the wing interior hollow, except for ribs and spars.

The four Fowler flaps add 45 square feet of wing area when extended and, in addition, increase the maximum lift capabilities of the wings well beyond that due to just the increase of area. This unique configuration would permit small heavily loaded wings for high speed cruise, yet slow, safe, touchdown speeds and short landing runs when fully extended. Partial extension would reduce takeoff distances and speeds, as well.

Rutan has put his research developments to good use. Fowler flaps are included on both fore and aft wings of his superb model 78-1 commuter design, and on the aft wings of Beechcraft's beautiful Starship I—a canard obviously reflecting Rutans influence in both design and construction.

Both will be reviewed later in this book under "Future Canard Developments."

Quickie, Model No. 54

Two men spent several years evaluating small engines searching for one with the following specifications for a small, light aircraft.

Fig. 3-19. The RAF Model 54 Quickie startled the modern aviation world with its introduction in 1978. The 240 pound aircraft achieved an 80 mph economy cruise while obtaining 100 mpg. The composite tandem wing featured composite construction throughout. It is marketed by Quickie Aircraft Corporation.

- 12 to 25 horsepower
- Light Weight
- Small Size
- Low Fuel Consumption
- Very, Very Reliable

They surveyed two-stroke engines, rotary engines, four-stroke motorcycle engines, Volkswagon engines and finally, industrial engines. All of them, except an industrial unit, failed to meet the criteria that they had laid down.

Starting their search in early 1975, Gene Sheehan and Tom Jewett, both development engineers, finally found the Onan industrial engine, of which more than one million had been produced for applications ranging from electrical generators to snow plows. This engine, in the aluminum version, modified for airplane use, could be reduced in weight to just over 70 pounds dry. It produced 18 horsepower at 3,600 rpm in a two cylinder, horizontally opposed four stroke with a time before overhaul (TBO) at 1,000 hours.

After much testing and modification in the areas of induction, exhaust, cooling, ignition system and mounting, they arrived at a suitable engine; much modified from the original.

Both met with Burt Rutan and suggested that he design a scaled-down single place VariEze, light enough and aerodynamically efficient enough to permit adequate performance with the Onan engine. Rutan sketched his Model No. 49 which he rejected due to the very low Reynolds number and high drag particularly of fins and canard wing, and because variations in the pilot's weight would result in excessive CG travel.

After several configurations, Rutan hit on the Quickie design. This is a tandem wing airplane—the almost equal distribution of wing area between fore and aft wings overcame the Reynolds number problem of the canard, while the large single fin did the same. The pilot's position was almost on the center of gravity, bringing the CG travel within acceptable limits—the pilot sits on the fuel tank. This model offered the weight, stall resistance and performance potential the three were seeking.

Busy with other projects, Rutan did not want to involve himself in marketing the Quickie kit. The other two did, so an agreement was arrived at that Rutan would fund, develop and test the Quickie.

He would be repaid out of future sales of Quickie and Quickie related designs, but have no involvement in marketing the Quickie.

The prototype flew reasonably well on initial testing in November 1977. It required a new canard configuration and airfoil and span changes on the aft wing.

Then followed an extensive five months of testing, involving 150 hours of flight, covering:

- Performance, stability and control throughout the center of gravity range at gross weight, over-design.
- Flutter tests to 162 mph indicated airspeed at 6,000 feet altitude.
- Fuel consumption measurements. At maximum cruise of 121 mph it achieved 80 mpg, and at economy cruise, 100 mpg.
- Stall, departure, and spin testing indicated the airplane could not be induced to spin at any time in the test program.
- Engine and system reliability. No power plant failures occured and nothing more than normal engine maintenance was needed.
- Landing gear energy absorption tests to FAA standards. This was reassuring in view of the unusual wheel placement at the foreplane tips, which would impose high bending loads at the canard root.
- Static load testing of the whole airframe to FAA utility category standards.
- Crosswind and turbulence testing. The airplane was flown in 57 mph winds.
- Independent pilot evaluations were successful.

The Quickie has plain flaps on the foreplane for pitch control. Inboard ailerons are on the aft-plane, and a fin and rudder are mounted at the end of a down swept tail cone with a steerable tail wheel on a tubular extension. The main wheels are enclosed in canard wing tip wheel pants of pleasing appearance. Braking of main wheels is independent for ground maneuvers. Due to the load on the front pair of wheels, the canard flexes. To avoid binding of the pitch control flap, special hinging was required.

The Quickie landing gear is very unusual. A taildragger normally relies on the elevators to raise the tail at the start of the takeoff run and then lower it, increasing the wing angle of attack for lift off.

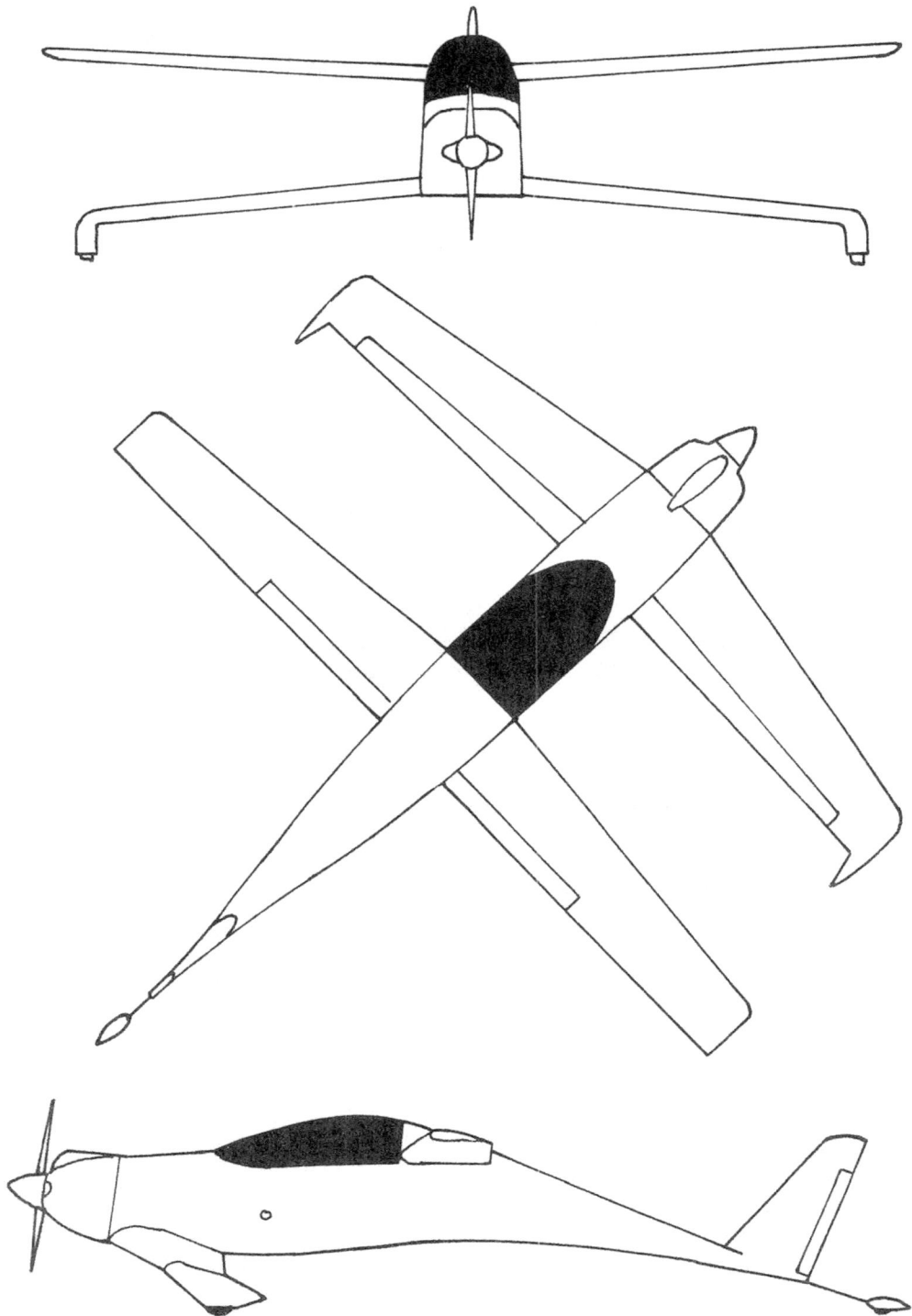

Fig. 3-20. Three view drawing of the Quickie.

There is just no way the Quickie canard elevators can operate as described above. Instead, the whole aircraft lifts in an almost level attitude as the takeoff speed of 53 mph is achieved. Climb at sea level is 425 fpm, and the ceiling is 12,300 feet. Ground effect, particularly on the foreplane, must be a big factor in cushioning takeoffs and landings.

Like other Rutan designs, the Quickie won the Outstanding New Design Award from EAA at the 1978 EAA Oshkosh, Wisconsin fly-in. Quickie Aircraft Corporation sells complete Quickie kits including the 42 inch diameter propeller and Onan engine.

This airplane has only basic instruments and limited NavCom equipment. It is a fun, safe, rugged little airplane that lands at a modest 59 mph. It is not aerobatic and the engine must be started by hand propping. Ground maneuvers are like race car driving.

One reporter is quoted to have said: "Flying the Quickie is the most fun you can have in the daytime, in public, without being arrested."

It should be noted that it weighs only 240 pounds empty—just a few pounds more than many large pilots, yet it achieves just under seven mph per horsepower. Larger, more powerful aircraft achieving one mph per horsepower, are considered efficient.

Burt Rutan has no involvement with Quickie Aircraft Corporation.

Amsoil Biplane Racer

Rutan designed this aircraft to racing biplane standards, which include 75 square feet of wing area and interplane struts, for Dan Mortenson and his sponsor the Amsoil Corporation, so that both could promote sale of a new synthetic aviation oil Amsoil is marketing. This beautiful airplane is similar to the smaller Quickie in configuration. It is powered with an IO-320 Lycoming engine of 160 horsepower, turning a fixed pitch propeller.

The cowling is the usual Rutan updraft cooling with an oval cooling air intake below the spinner and vented on the top surface. A lower carb-air intake is provided.

Since the fuselage behind the engine narrows down considerably, large cheek fairings streamline the engine cylinders and contain 7½ gallons of fuel each. Another 16 gallons is contained in a tank behind the pilot. The pilot's enclosure is narrower than the fuselage and extends to form a spine that blends in with the Dorsal fin. The fuselage overall is 22 feet long.

Wing sections were computer developed for high speed (200 mph), 90° banks. The foreplane, of 20 feet - 5 inches span, is covered in carbon fibre instead of fiberglass, to support the landing loads imposed by the big engine. The moldless composite system employs PVC foam as the core material.

The aft wing has its ailerons close to the wing tips of the 22 foot span surface. Curved biplane struts are designed to resist flutter and to flex with the front wing as it flexes, particularly on landing.

The aileron linkage was changed from aluminum to steel tubing to improve strength, while the aileron travel was increased from 6° to 12° with a

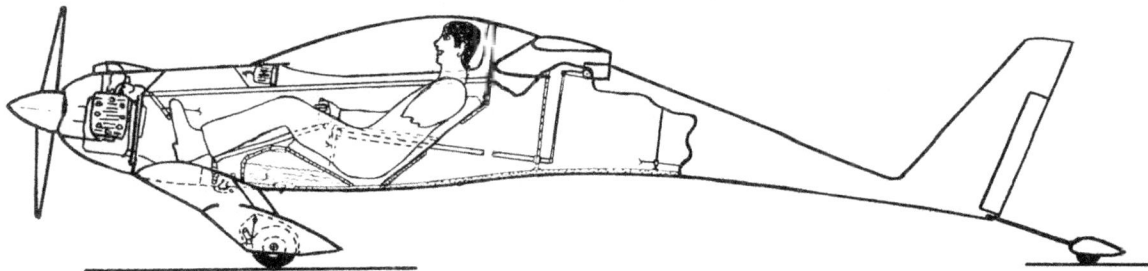

Fig. 3-21. Inboard profile of the original Quickie reveals simplicity of the configuration and its main components.

Fig. 3-22. The RAF Model 68 Amsoil Biplane Racer was designed to the specification for racing biplanes. Designed by RAF in 1979, it was built by an Amsoil-sponsored team, and was first flown in 1981. It is similar to the Quickie, with the addition of a small horizontal tail, geared to the canard elevator, to provide optimization of canard camber for all racing flight conditions. Don Downie photo.

design criteria of 16°. Prior to the change, the airplane nearly crashed, after involvement with wind shear and prop wash.

The fin and rudder are carried on the tail cone, the tail wheel is steerable and mounted on a tubular extension that flexes for shock absorption.

The small flying "T" tail atop the vertical fin is geared to the canard's elevator. It is a trimming device that insures the elevator is at its most efficient position in turns (+2°) and on straight-away (-2°).

The engine is not up to full race standards. It has a starter and alternator and a modified exhaust system that exits from the bottom of the cowl, positioned so that hot exhaust gasses do not impinge on the canard. Cockpit visibility is superb.

NASA tests with a coating of sublimating chemical on the wings showed that the airfoil had extensive, natural, laminar flow averaging 60% of the wing chord. It even persisted in the inboard areas affected by the propeller slipstream.

The racing propeller provides a top speed of 232 mph. The 500 x 5 wheels and Cleveland brakes comply with NAA standards for biplane racers.

Empty weight is 854 pounds, while gross is 1,167 pounds. The wing sections are designed for racing, and if the airplane is slowed below 70 mph, the plane will suddenly drop a wing without warning. It is a highly efficient racing machine and must be flown accordingly. No slow approaches are permitted!

The contract between the two principals called for Rutan to do the designing and Mortensen to build two airplanes, with Rutan retaining the design rights.

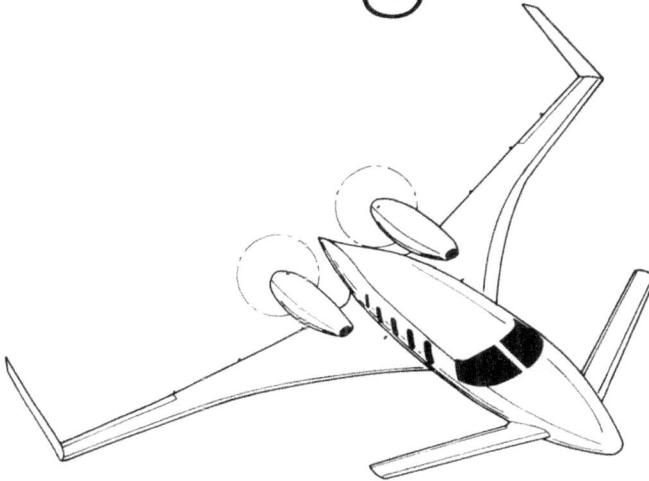

Other Amateur-Built Canards

Quickie 2

Quickie 2 is a two place development of the Quickie, produced by a colaboration between Quickie Aircraft Corporation, headed by Tom Jewett (unfortunately killed in a non-Quickie one-of-a-kind experimental aircraft crash) and Gene Sheehan, and Leg Air Aviation Ltd., headed by Canadian Garry LeGare.

They cooperated in the design, and LeGare built the prototype. The detailed design was completed in early 1980, and LeGare started his work promptly. The first flight test was on July 1, 1980, after passing the rigid static load testing and inspection requirements of the Canadian Department of Transport. In mid-August 1980, after Canadian flight testing, the Q2 was transported to Quickie Aircraft Corporation in Mojave, CA to complete the flight test program and static load testing.

The engine selected for Quickie 2 was the proven Revmaster 2100-D (converted Volkswagen) detuned to just under 65 horsepower (from a turbo charged unit of 85 horsepower) to obtain a reliable, rather than powerful unit.

Similar to the Quickie externally, it is a two place, side-by-side seating aircraft of high performance on low horsepower. However, it is not just a scaled-up Quickie—all its aerodynamics and structure represent a new design.

The desired CG envelope was fully explored and achieved, and the design gross weight reached. With full electrical system (and starter) and radio, the empty weight came to 537 pounds, and it proved capable of 180 mph. Engineering evaluation proved Quickie 2 to have an "equivalent flat-plate drag area" (a method of comparing aircraft drag by equating it to a flat plate of equal drag, in an airstream) of 1.37 square feet. This compares to other two-placers of flat plate areas in the 1.4 to 1.7 square feet range.

Specifications of Q2 follows:
- Spans — wing: 16 feet - 8 inches
  canard: 16 feet - 8 inches
- Total wing area — 66 square feet,
  wing loading 15 pounds per square
  foot

Fig. 3-23. The Quickie 2/200 is a development of the original Quickie. It was a result of a colaboration between Quickie Aircraft and Canadian Garry LeGare. The 2 version was powered by the Revmaster 2100-D Volkswagen conversion, while the 200 is a modification designed to mount the Continental O-200. The later version featured a top speed in excess of 200 mph, and a climb rate of over 1,200 fpm.

- Weights — empty 537 pounds
  gross 1,000 pounds
  useful load 463 pounds (86% of net weight)
- Climb — 800 rpm (pilot and passenger)
- Top speed — 180 mph
- Landing speed — 70 mph (pilot and passenger)
- Ceiling — 15,000 feet (pilot and passenger)
- Fuel consumption — 44 mpg at maximum cruise, 60 mpg at economy cruise
- Power loading — 15.6 pounds per horsepower (st 64 horsepower)

Gross weight performance of the Quickie 2 and of the VariEze first prototype with the VW engine are very close, as are wing and power loadings.

Appearance-wise, the departure from Quickie consists of a wider fuselage, spinnered propellers, conventional twin cooling air intakes in line with the propeller shaft and down draft cooling, exiting on the fuselage bottom ahead of the firewall. The pilot sits on the fuel tank which is close to the CG, so that fuel burn off has little effect on trim.

The prototype threw one blade of an experimental metal variable pitch propeller and the resulting forced (and forceful) landing destroyed the airplane. Fortunately pilot Sheehan was uninjured. A second prototype won the first Cafe 250 race by an impressive margin. The design has the stall/spin and departure stall-free characteristics of the Quickie: but in common with the VariEze, has quite a high landing speed, calling for a long slow approach and landing flare, right

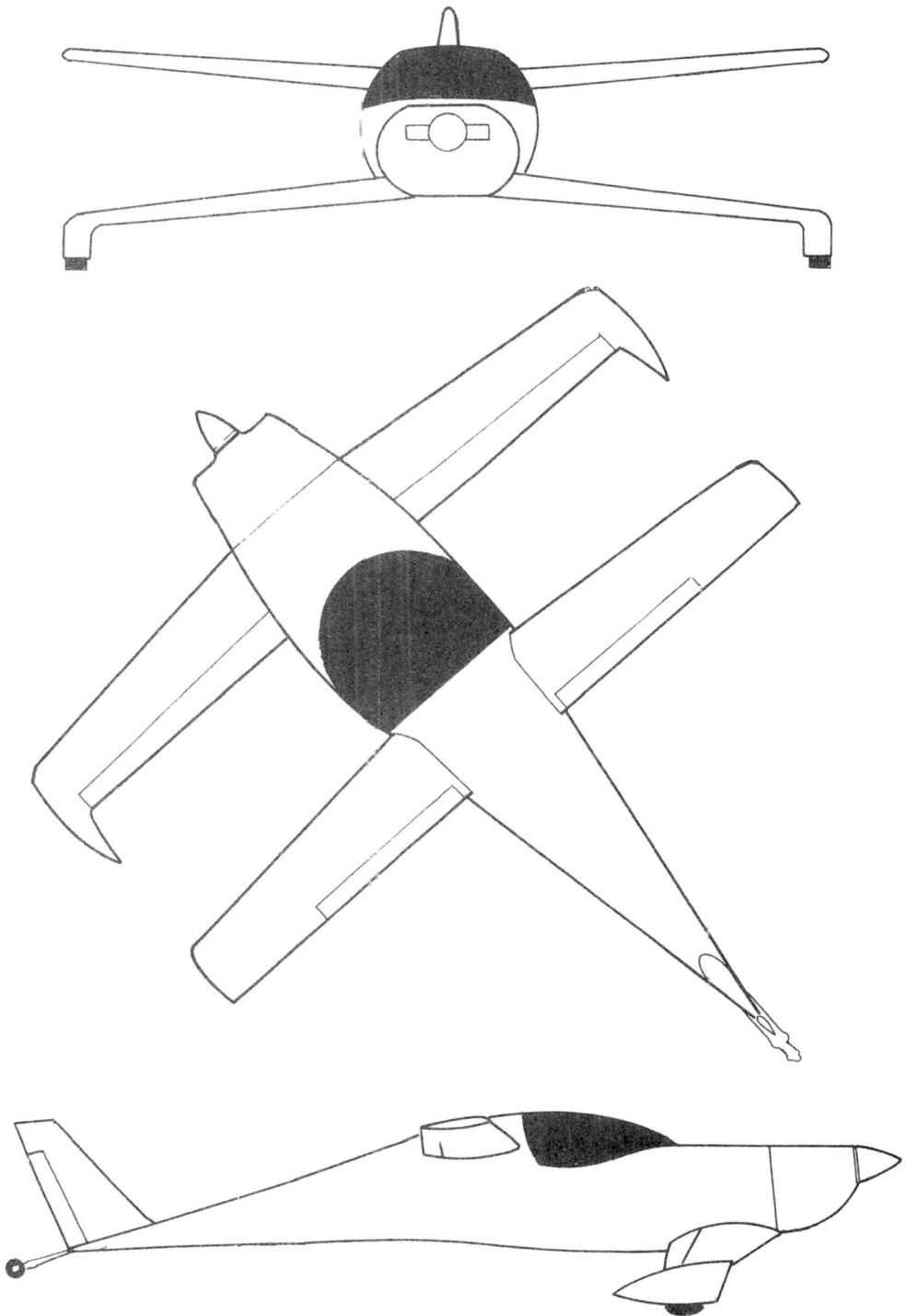

Fig. 3-24. Three view drawing of the Quickie 2/200.

on the runway. Stall landings from one foot can be "very exciting."

It is said that "emulation is the sincerest form of flattery," and Burt Rutan has certainly received his share. His developments have lead to a large and complimentary number of tail-first aircraft designs, some of which are described in the following pages.

> In the October 1981 issue of The Experimental Association's monthly magazine *Sport Aviation,* Rutan wrote: "Aero elastic effects (airframe twisting and bending with speed and angle of attack) which are normally negligible in a conventional aircraft can change a stable airframe into dangerous instability at high speeds. Conversely, aeroelastic effects can aid high speed stability with proper structural and aerodynamic design."

> "The designers' data base for these types of designs (canards and tandem wings—author) is extremely limited, and the importance of understanding their aerodynamics is great. A strong possibility exists for the introduction of inferior designs that do not meet the requirements for satisfactory flying qualities and safety, thus giving the *configuration* a bad reputation. I am proud of the potential promised by this configuration to improve aviation safety and efficiency. Thus, I encourage other developers to not accept less than perfection in their aircraft that are offered to the public. *I am willing to assist in solving aerodynamic or aeroelastic problems during your flight test programs if you are willing to incorporate the indicated fixes. Let's work together for a better aircraft rather than separately in a relative vacuum."*

Happily, several of the designers of the aircraft described below have accepted Rutan's generous offer and have had their aircrafts' characteristics run through his computer program. Many others have also sought his advice and suggestions.

Any reader with ambitions to develop and fly canards would be well advised to consult Rutan, before getting into detail design.

As a note of interest the Gyroflug speed canard, a Rutan VariEze look-alike, has been certified in its native West Germany. It is powered with a 110 horsepower Lycoming O-235, driving a three blade Hoffman constant speed prop which is alleged to improve its takeoff vis-a-vis the VariEze and which, in fine pitch, produces drag for steeper approaches. Components will be produced in molds, at the rate of 50 units per year.

Quickie 200

The Quickie 2's VW engine was never designed to be an aircraft engine. Its problem is that it is designed for efficient power output at rpm's much higher than the range for efficient propeller operations. Its high power output is compromised by propeller losses.

It was desired to use the Continental O-200 of 100 horsepower, but higher weight. This added weight would have resulted in an unacceptable canard wing loading, making for very high over-the-fence landing speeds.

Another point was that the canard airfoil, a laminar flow section, suffered considerable lift degradation in rain, or if the leading edge was contaminated with insect carcases or grass cuttings. Some Quickie and Quickie 2 pilots could hardly maintain altitude, at full aft stick, in heavy rain, making landings hazardous. One Quickie 2 builder decorated the full length of both wing and canard leading edges with trim strips .005 inches thick—just enough to trip the laminar flow, limiting both canard lift and aileron control. After burning off all his fuel he made a hard bouncing landing. Since the Quickie 2 canard carries 65% of the aircraft weight, the trim strips caused a nose heavy condition. Sanding off the strips cured the problem.

A new canard with NASA designed section LS(1) - 0417 MOD was installed which, it is claimed, eliminates trim changes in rain. It also expands the CG range, lowers maximum gross weight stall speed, provides shorter takeoff and landing rolls, and reduces rudder sensitivity as well as "inhibiting spin tendencies."

Top speed is now over 200 mph, with climb at 1,200 fpm. On takeoff, the aft wing lifts first—due to its lower wing loading, identified by absence of tail wheel rumble. One oddity occurs however, in flight; the air flow on the exposed lower portion of the main wheels causes them to rotate even faster

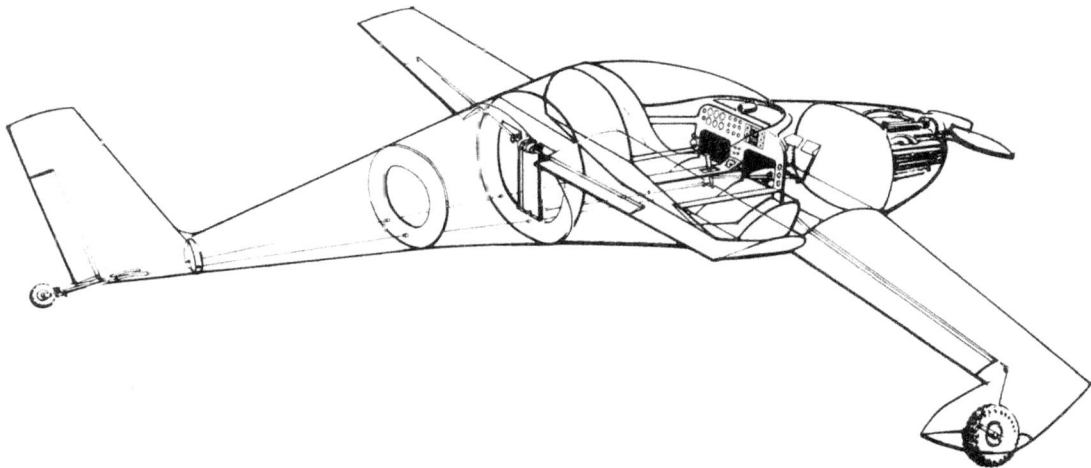

Fig. 3-25. Cutaway of the Quickie 2/200 reveals its major components. It is constructed of foam/fiberglass composites.

than after lift off. Any unbalance in those wheels causes an unexpected shuddering of the whole aircraft that requires light brake application to arrest. The author has experienced this "shudder" in a Cessna, after a high speed "unstick," necessitating a tap on the toe-brakes to stop it, but never in flight. The Quickie pilot must release his brakes before touching down—particularly if a cross-wind calls for one wheel contact. Because of its aerodynamic cleanliness, it lands even faster than Q2 and calls for care.

Another feature is ailerons that can be "reflexed." These are mounted inboard on the aft wing and can be adjusted up or down together, yet retain the up-down action of normal ailerons. This same feature was used on the Vari-Viggen. It was intended to permit more nose-up trim in rain—but could also be useful as lift increasing flaps, in conjunction with some back-stick for longitudinal trim, permitting lower landing speeds and shorter takeoff runs.

Both Q2 and Q200 are offered as complete kits of premolded components, less the engine. They can be fitted with 2 different VW engines, one normally aspirated—the other turbo charged, and of course, the Continental O-200.

The broader fuselage (compared to the single place Quickie) provides a larger instrument panel so that full IFR instruments and NavCom equipment can be installed—as well as dual controls.

THE VIKING DRAGONFLY

This aircraft was developed almost simultaneously with the Q2, but completely independently. In outward appearance they are remarkably similar. It is alleged that the "Dragonfly" is a scaled up version of the single place Quickie, and utilizes the airfoils and aerodynamics of that small aircraft, but it isn't so.

Viking Aircraft Limited of Eloy, Arizona is the kit supplier, and is under control of the Taylor family. Rutan was not involved in the Dragonfly design, or testing.

It is interesting to note that FAA regulations state that in order for an airplane to qualify as a homebuilt, the builder must do 51% of the work and the FAA evaluates the high volume designs to insure compliance with these regulations. The airplane itself is subject to FAA inspection as building progresses.

Despite their similarity, the Dragonfly and Quickie 2 are very different aircraft as the figures below indicate:

	Dragonfly	Quickie 2
• Total Wing Area Sq. Ft.	97	66
• Gross Weight Pound	1075	960
• Horsepower	45	64
	(1600cc VW)	(2100 Revmaster)
• Wing Loading/Sq. Ft.	11.08	14.54
• Power Loading/HP	23.8	15

The lower wing loading and higher power loading of the Dragonfly render it slower, with an estimated top speed of 150 mph, but this also makes for a slower touchdown speed. With presumably, the 1835cc VW HAPI conversion as power, the latest performance figures quoted in Dragonfly advertisements call for a 400 pound payload, a cruise of 165 mph, a climb rate of 850 fpm, and a stall speed of 52 mph, with landing at around 57 mph. The larger engine would have an electrical starter and alternator. Full IFR instrumentation could then be provided, along with dual controls.

A pilot's report on the Dragonfly indicates easy lift off, but landings calling for precise speed control to avoid "floating" in ground effect over the runway. An extra 10 mph calls for an extra 1,000 feet of runway, due to the aerodynamic cleanliness of this airplane. Pilots are warned to *fly* the aircraft all the way to the ground, not to stall it on from more than one foot above ground level (AGL).

The canard is very springy, as is the tail wheel mount. Stalling on from three to four feet, results in a bounce from main to tail wheel and back again that could be damaging to the propeller tips. Bounding down the runway like a "gazelle-with-its-tail-on-fire" is not a dignified nor graceful way to land. The Quickie 2 and 200 must have similar characteristics at their higher landing speeds.

Externally, the major points of difference are

Fig. 3-26. The Viking Dragonfly, while very similar in appearance to the Quickie 2/200, is a completely different aircraft. It was developed entirely independently of the Quickie, and without any knowledge of it. It is a larger, slower aircraft, and as such, probably more accommodating to the low time pilot.

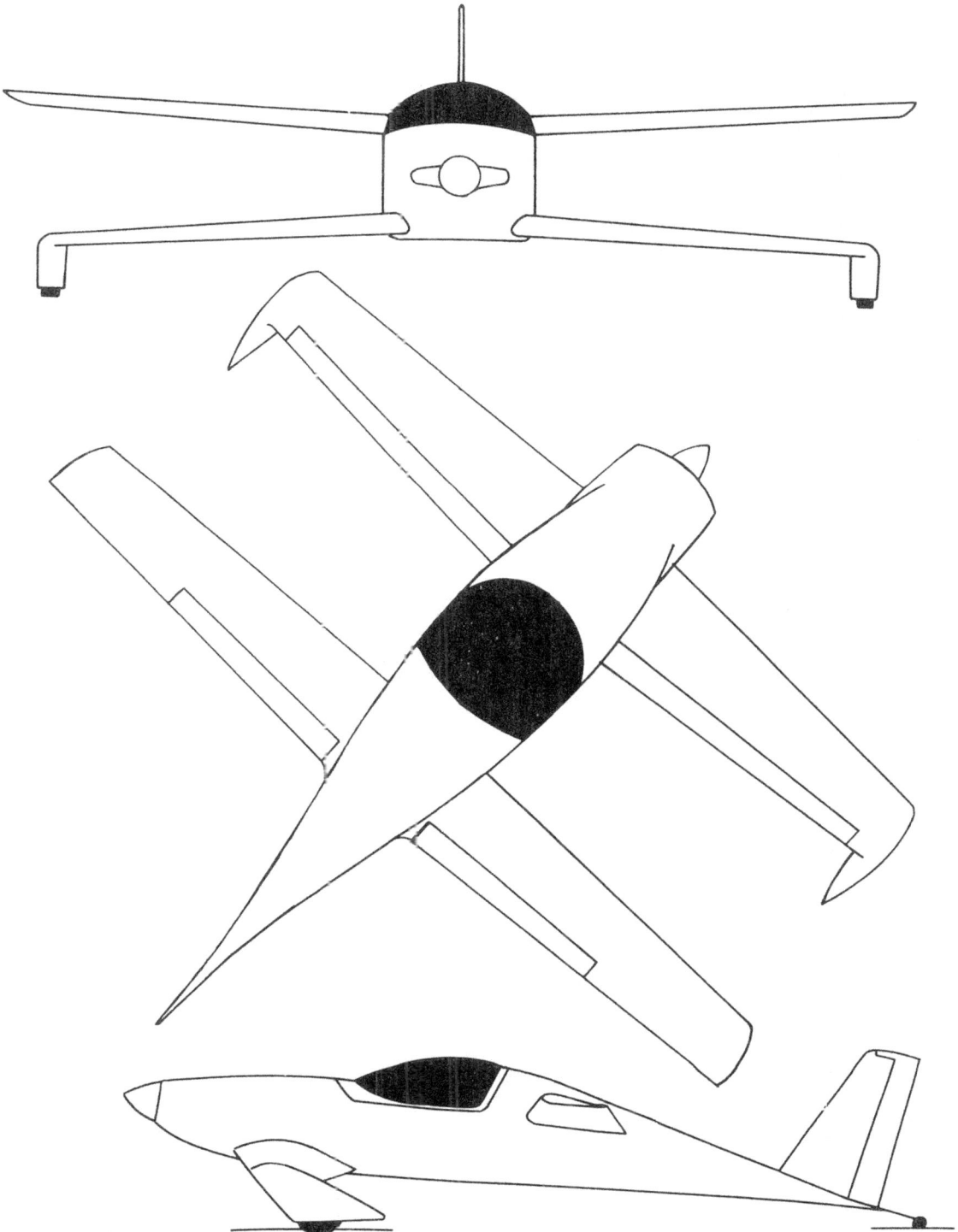

Fig. 3-27. Three view drawing of the Viking Dragonfly.

the lower thrust line, with the fuselage nose arranged almost symetrically above and below it, and with cheek cowls for fairing the engine cylinders. Cooling air baffling is designed for minimum drag.

Vortex dispersing aft wing tips add grace to this surface, while ailerons are inboard for easy control linkage. The rudder is mass balanced by a shielded horn. Small "servo" surfaces behind the foreplane elevators eliminate the bungee springs of the Quickie and permit better elevator control feel. The foreplane has a modified GU airfoil section for more gentle stall.

Like the Quickie, Quickie 2 and 200, the two almost equal area wings permit high aspect ratios yet low span. Overall, the airplane is very compact. The side-by-side seating and fuel tank location under the seats, both almost on the CG, reduce CG shift with variations in pilot weight and fuel consumption. The fuselage maximum width is 43 inches.

Both wing and canard are of one piece construction. The foreplane's spar cap is of carbon fiber for strength, due to its length (20 feet), the aircraft's greater weight (1,175 pounds) and tip mounted wheels, imposing large canard root behind loads.

Dual controls and an instrument panel large enough for full IFR and NavCom equipment are provided, making flight instruction possible.

Since the foreplane carries more than 60% of the aircraft's weight in flight, the elevators naturally want to lift under airloads, hence the small servo surfaces or "tabs" referred to above. Spring systems on both elevator and ailerons take care of trim changes. No rudder trim is needed. The tail wheel is mounted at the end of a flexible tubular fuselage extension, and is steerable through linkages to the rudder.

Construction is foam/fiberglass, but utilizes a scored foam that is flexible for compound curves. The foam is 4½ pound density water blown polyester based urethane.

The "Dragonfly" won the outstanding New Design Award at the 1980 Experimental Aircraft Association Oshkosh Convention.

As one who is familiar with Canadian and northern United States winters, the use of close fitting main wheel pants of the Quickie/Dragonfly type, or any other type, during winter months is a "no-no." Slush and/or snow will pack into the wheel cavities and freeze in flight, making for very effective brakes. Current practice is to remove such fairings for winter flying.

The Dragonfly retains the departure stall and stall/spin freedom of its configuration. Full aft stick calls for a low amplitude/time cycle bobbing, while power can be used for ascent or descent. The bobbing is caused by the foreplane partially stalling, sinking slightly to a lower angle of attack and unstalling, to repeat the process.

The aft wing cannot be stalled, and full aileron control is thus always in effect, even in steeply banked turns. Centrifugal force only hastens the foreplane stall, limiting the turn radius. With its lower approach and landing speeds, the Dragonfly would seem more suited to the low-time pilot, although some form of drag device for steep approaches and less floating in ground effect would seem desirable.

THE PUFFER COZY

The Cozy is a side-by-side seating variation of Rutan's Long-EZ, designed by Nathan D. Puffer, with Burt Rutan's knowledge and consent. The design was run through the Rutan's computer program, and utilizes the Long-EZ aft wing, ailerons and winglets.

The engine is an O-235 L2C Lycoming of 188 horsepower in the pusher configuration. Span is 26.1 feet, wing area 95.6 square feet, gross weight 1,500 pounds and fuel capacity 52 pounds. At 75% power, its range is 1,200 miles at 180 mph. At 40% power, range is 1,800 miles at 143 mph. The Cozy provides fast, long range and fuel efficient transportation for two people and reasonable luggage, which is better than most two-place commercially available airplanes.

The engine requires a mechanical fuel pump, and has an alternator and electric starter. Flight controls are the Rutan side-sticks. Other controls are located in a central console accessible to both occupants. A large area instrument panel provides space for radios and full IFR instrumentation of standard size.

With the side-by-side seating, pilot and passenger weight totalling 340 pounds is no problem.

Fig. 3-28. The Puffer COZY is a side-by-side seating version of Rutan's Long-Ez. It was designed by Nathan D. Puffer, with Burt Rutan's knowledge and consent, and was run through the Rutan computer. It cruises 180 mph on 75% power, while offering a 1,200 mile range.

Lighter weight or solo flight, calls for a weight and balance check and possible addition of ballast in the nose compartment. The battery can be moved from the firewall to the nose. A back seat for one small person or luggage is included, as well.

The wing loading at gross is 15.7 pounds per square foot, while power loading is 12.7 pounds per horsepower. Approach and landing speed at gross is a high 80 mph. A spring loaded "set-it-and-forget-it" belly mounted air brake is used to steepen approach to landing.

The landing gear is tricycle, with a retractable nose wheel that is free castering. Ground maneuvering is done with differential braking. Both rudders may be deflected outward simultaneously

for additional drag on landing. In flight, only one rudder is used, that on the inside of the turn, but its use is infrequent except for crosswind landings or takeoffs.

The Cozy shares the safe flight characteristics of Rutan's designs. There is no divergence from controlled flight—no stall/spin. The canard airfoil is angle of attack limiting so that, in level or turning flight, the angle of attack of the rear wing cannot reach its stalling angle. Full aft stick, engine idling, results in a stable high sink rate, which is quickly halted by an increase in engine output.

Canards are basically more efficient. Both wings contribute to lift—permitting a smaller aft wing with lower bending loads and lighter weight. Drag is reduced, and better performance with

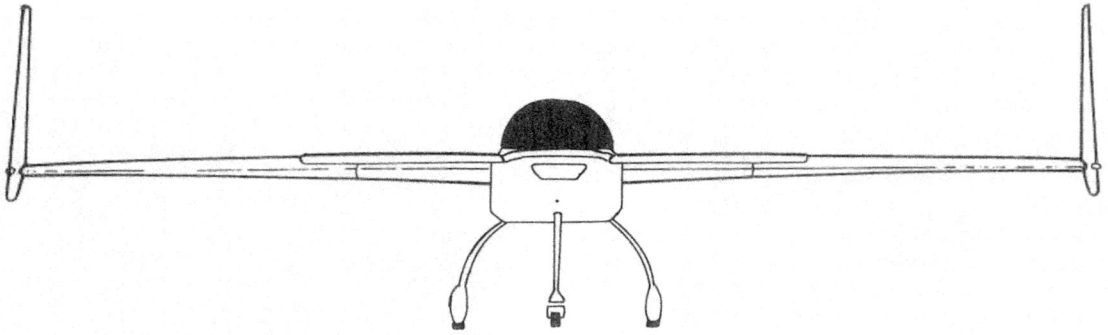

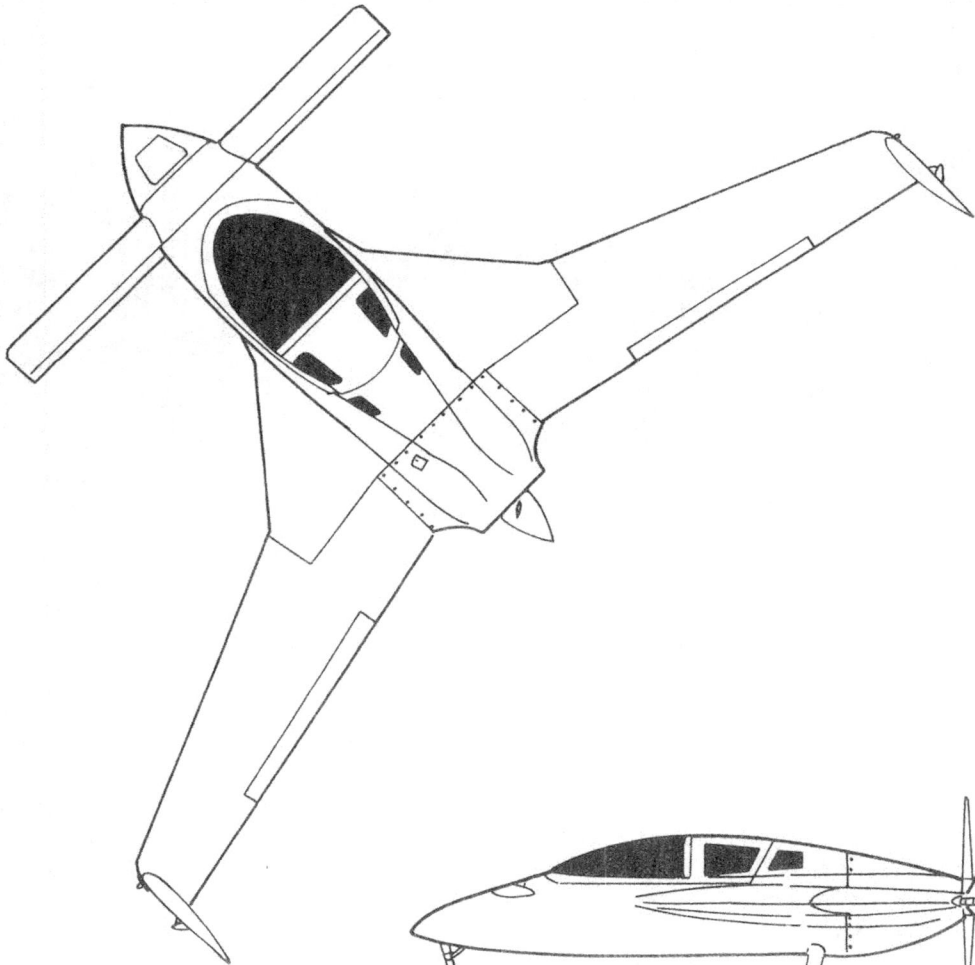

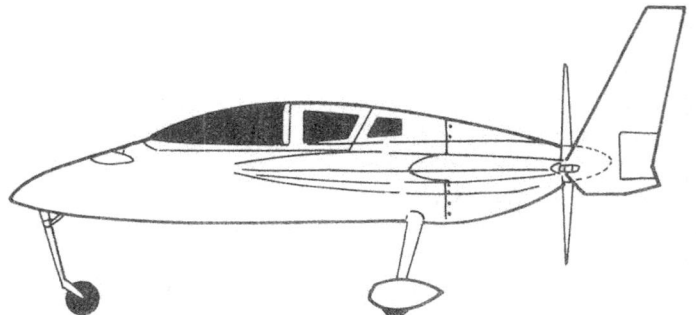

Fig. 3-29. Three view drawing of the Puffer COZY.

less horsepower results. When will the "big-plane" manufacturers like Cessna and Piper react to these facts.

The Cozy flies well. It is solid and stable with good aileron response, hands-off stability, and because of its relatively high wing loading, it is good in turbulent air. Forward visibility is excellent.

Landings are "jetliner" like. A *stabilized* final approach is established, and a slight pitch adjustment stops the rate of descent. "Frozen" in this attitude, and as speed bleeds off in ground effect, the plane will set itself firmly on the runway. Held off too long, it may settle too solidly. The *stabilized speed* and *attitude* on approach is all important.

The Cozy incorporates the same University of Glasgow airfoil in the canard and an Eppler airfoil in the main wing. Rain effects a slight pitch change on the canard, but is insufficient to compromise performance or safety.

The Cozy was completed and first flown on July 19, 1982. By exclusive license from Rutan, the CO-Z Development Corp. of St. Paul, Minnesota is offering the Cozy as a homebuilt.

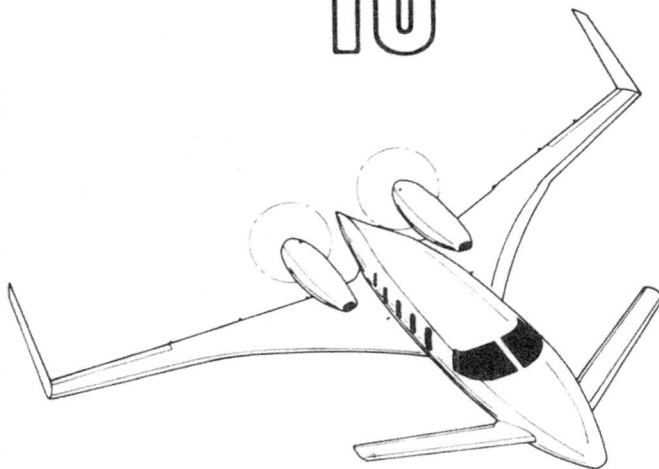

10

Ultralight And Very-Light Canards

The Ultralight movement has spawned some weird and wonderful aircraft; many reminiscent of the planes of the early pioneers, like the Wright brothers.

Some are very crude; the pilot is fully exposed as are engines, framework and landing gear; designs vary widely, some are tailless, some have inverted "V" tails—others conventional tails; still others are canards. Engines are pusher or tractor, some belt driven. Some have twin engines and would you believe, "biplane" propellers—one prop mounted directly ahead of the other on the same shaft, rotating as one.

Some interesting canard water planes are also flying. This section will deal with four designs, two of them watercraft.

RIVERS R-I

This is an intriguing little single-place canard which incorporates some unusual features. Designed by John McReynolds, a design engineer on the Lockheed C5A; built by Eddie Clark and financed by Rivers Stone of Greenville, South Carolina, it displayed beautiful workmanship at the E.A.A. Oshkosh 1983 Convention.

Both wings are rectangular in plan form; the foreplane is fitted with full span trailing edge flaps as elevators, and is of a cambered section similar to NACA 4415. It sits low at the aircraft's nose, and the pilot's view is not interrupted by it to any great extent.

The aft wing has no dihedral, as does the foreplane. Its section is semi-symetrical, and it is provided with conventional ailerons.

The shapely canopy runs from the aft wing leading edge to approximately mid-point on the canard. The helicopter-like centrally mounted instrument panel permits excellent visibility. All four wing tips are simply squared off.

The unusual feature is a large vertical fin and rudder, also rectangular in outline. It is mounted on a short tubular extension of the upper fuselage. The propeller hub forms part of this boom, and rotates just behind the aft wing trailing edge. The hub has a large diameter bearing mounted

Fig. 3-30 The Rivers R-1 is an all composite, single-seat, ARV type very light aircraft. It features a unique prop on boom drive system.

inside, and is powered by a two-cycle engine with belt drive. The engine appears to be wholly enclosed in the aft fuselage below the wing NACA type cooling air inlets.

The pilot is located just ahead of the CG, and probably sits on the fuel tank to minimize CG shift as fuel is burned-off. The landing gear is tricycle with straight legs running from the fuselage close to the wing leading edge, with the nose wheel fork-mounted on a tubular strut, but non-retractable.

At the time of its Oshkosh display, this little plane had flown in ground effect only. Extensive production tooling was completed just before the convention for this fascinating concept. The FAA has however, imposed both slow and top speed limitations for ultralights. The Rivers R-1 cannot fly slowly enough to meet the low speed criteria. The project has been abandoned, and is for sale.

BEZZOLA - RETRO GB2

This is a homebuilt designed and built by Gion Bezzola of Switzerland, a pilot by profession. It was inspired by the famous Rutan VariEze.

The Rutan has a span of 18.37 feet, overall length of 15.09 feet and a gross weight of 904 pounds. It is powered by a 1600cc VW conversion, developing 45 horsepower at 3,600 rpm, with single ignition and carburetor heat. It drives a Bezzola designed 51.2 inch diameter, three bladed propeller made of Kevlar over a foam core, with ground adjustable pitch. The engine is mounted under the aft wing trailing edge and neatly cowled with "arm-pit" air intakes.

The landing gear is tricycle, made of fiberglass, wood and metal, with wheel pants on the mains, and hydraulic disc brakes. The non-retractable nose gear utilizes the trailing beam principal. Taxi tests indicated the need for a shimmy damper

The aft wing has a straight trailing edge and swept back leading edge, as does the canard, whic set very low on the fuselage front. Wings, winglets and canard were built of fiberglass, epoxy and foam. Conventional ailerons, and canard trailing edge elevators provide pitch and roll control. The winglets incorporate rudders for yaw control.

The fuselage is constructed of square section

aluminum tubing, a covering of aircraft plywood riveted to the tubing and skinned with fiberglass and epoxy. The angular canopy was made of flat sheets of acryl-glass glued to a framework of light steel tubing/foam/fiberglass. With the pedestal mounted instrument panel similar to that of a helicopter, visibility is superb, with no interference from the low mounted canard.

Flight testing included landing and takeoff run determination, climb speed, stall speed, stability in all axis, effects of power variation on trim, behavior in turbulence, and crosswind handling during takeoff and landing.

Modifications found necessary were: reduced canard angle of incidence, improved control linkage, better balance between elevator and aileron travel, improved directional control, a new and less noisy propeller, and as mentioned, a nose wheel shimmy damper. Most noticeable are the enlarged winglets.

Empty weight is 639.45 pounds, fuel capacity 12.15 gallons, fuel consumption 2.9 gph at 3,400 rpm. Maximum cruise is 118 mph, rate of climb is 492 fpm, and stall speed is 59 mph. Touchdown is in the 65 mph range, a fairly modest speed.

SunRay Amphibian

The SunRay, a product of the Sun Aerospace Corporation of Nappanee, Indiana is a canard amphibious very light aircraft.

It features rigid wing construction and conventional three-axis controls. The airframe is built primarily of Kevlar and is available with an open or enclosed cockpit. High performance laminar airfoils, sidearm hand controller, retractable tricycle landing gear, differential main wheel brakes and limited castering nosewheel, are features.

Dimensionally, the SunRay is 32 feet in span, 13 feet - 4 inches long and has a wing area of 126 square feet.

Fig. 3-31. The Bezzola Retro GB2 is a single-seater designed by Gion Bezzola, a Swiss pilot. It was inspired by the Rutan VariEze. Powered by a 45 hp VW conversion, it offers a cruise of 118 mph, and a stall of 59 mph.

Fig. 3-32. The SunRay is an amphibious composite construction single seater of unusual lines. The outboard wing panels are detachable, making for an eight foot wide unit suitable for transport over the road. It features a conventional three axis control system, and is powered by a converted snowmobile engine of 35 hp.

Gross weight is 510 pounds, while empty weight is 250 pounds. The useful load of 260 pounds is 104% of empty weight. Wing loading is 4.04 pounds per square feet of wing area.

Power is a 440-B Kawasaki of 30 horsepower. Maximum level speed is 63 mph; cruise 43.56 mph and stall is 27 mph, calling for a landing speed of about 32 mph. Ceiling is 13,500 feet, range on five gallons of fuel with 45 minutes reserve is 145 miles. Vne is 85 mph.

This amphibian is of unique configuration. The drawing does not detail the ailerons, but the mass balance projecting upward from the wing rear indicate conventional roll control. The canard has a slotted flap elevator. Yaw control is composed of rudders on the large fins.

On the water the plane floats on the hull, and on outboard floats in a form of tri-float arrangement.

The floats contain the retracted main wheels; the nosewheel retracts into the hull forebody. The power plant nests in the space below the inverted V of the wing center section, behind the pilot. The propeller shaft is set high, above the wing trailing edge and is belt driven from the engine.

The aft floats project forward in two booms running to a point beneath the canard. The pilot sits just ahead of the CG. Fuel is located in the hull afterbody behind the deep step in the hull.

The SunRay amphibian is a very interesting aircraft. Unfortunately, details of its flight characteristics and behavior in air and on water are lacking but it is assumed that it shares the safe habits of other well designed canards.

Detaching the outboard wing panels results in an eight foot wide trailerable unit. A two-place model is currently under development.

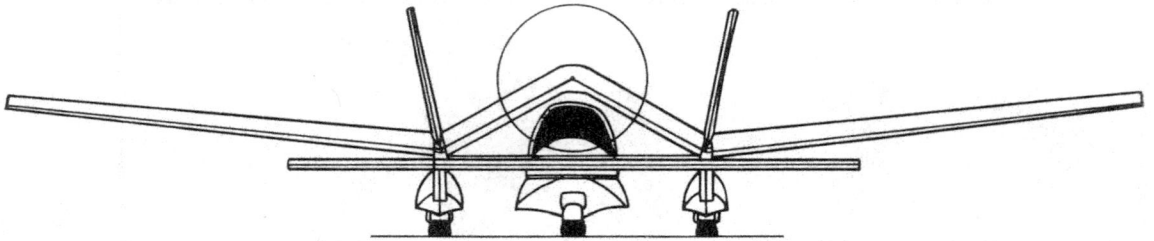

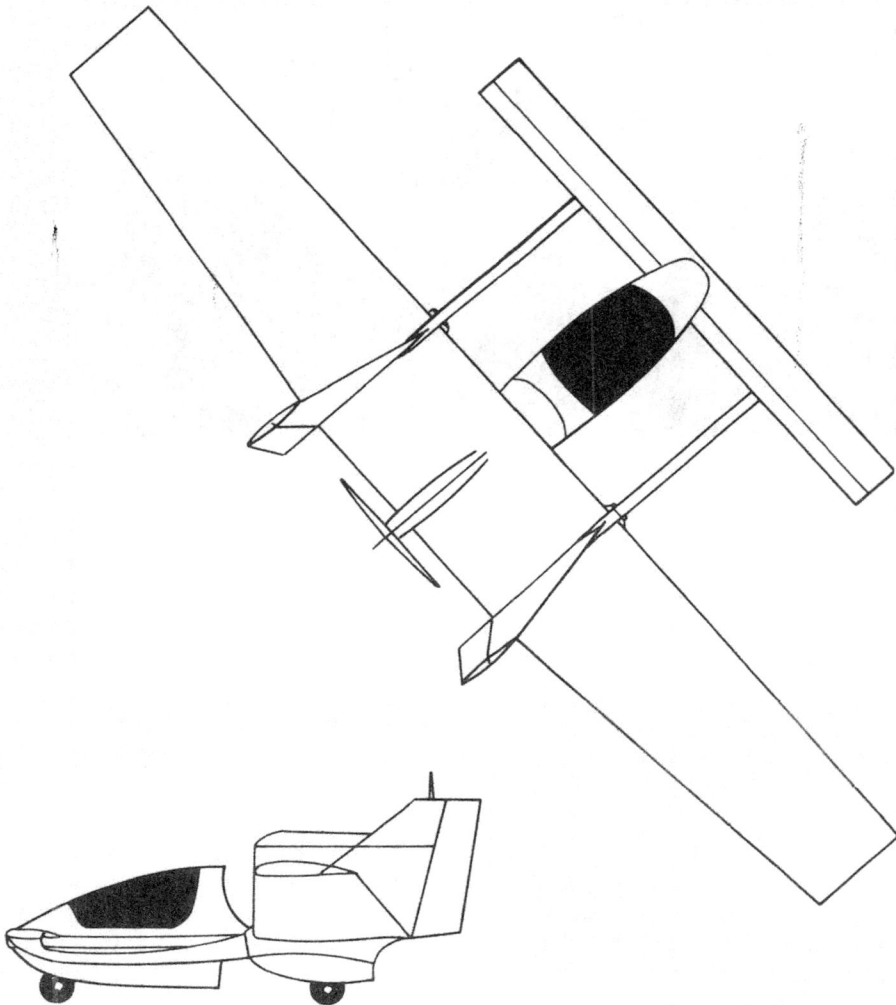

Fig. 3-33. Three view drawing of the SunRay amphibian.

Goldwing UL

FAA regulation FAR part 103 of October 1982 prevented the standard Goldwing from qualifying as an ultralight, due to exceeding the allowed limits of speed and weight. This requires registration of the aircraft and a student pilot solo permit. The standard homebuilt Goldwing kit has FAA approval as a "homebuilt" and qualifies under the "51% rule."

This writing concerns the Goldwing UL, which qualifies as an ultralight under the FAA regulations. While the photo and three view show the homebuilt version, the UL is very nearly identical; but it's factory built, not a kit.

The UL employs a cantilever wing, carbon graphite spar caps; Glasgow University high lift wing sections; NASA winglets, vacuum molded skins, hi-density foam cores and Dupont Kevlar.

Pitch control is via the canard trailing edge flaps. Roll control is achieved by small outboard ailerons, assisted by three spoilers on the wing top surface ahead of the aileron. They rise simultaneously and proportional to up-aileron action, and are claimed to be very effective. Directional control is provided by large patented split rudders. Deployment of both rudders simultaneously permits steeper approaches and reduced float in ground effect.

Wide stance main wheels and a steerable nose wheel make up the undercarriage. The engine is a 25 horsepower Zenoah with a 2.25 to 1 reduction unit, and is fully exposed.

Gross weight is 540 pounds, empty 250 pounds. The aircraft is 12 feet long, 30 feet in span and has a total wing area of 128 square feet. Wing loading at gross is 4.21 pounds per square foot. Power loading is 21.6 pounds per horsepower.

The UL cruises at 63 mph, with a "never exceed speed" of 70 mph. Stall speed is 26 mph, while landing occurs at 30 mph. Range at cruise is 100 miles and takeoff and landing rolls are 150 and 200 feet respectively.

The pilot sits on the CG in an open cockpit with a windshield, and is provided with an airspeed indicator, seat belt and shoulder harness. The aircraft has a useful load of 290 pounds or 116% of its empty weight, and carries three gallons of fuel. Controls are the conventional stick, with rudder pedals linked to the nosewheel for on-the-ground steering.

It has a well designed canard's stall/spin free characteristics, flies "hands-off," and is spirally stable.

The Goldwing has an interesting sister model, the Goldduster (not an Ultralight), capable of a useful load of 350 pounds. It carries a 15 gallon hopper for agri-spraying, with micronair fan driven spray heads. As such, it offers a low cost

Fig. 3-34. The Goldwing can be purchased either as an amateur-built ARV, or a factory-built ultralight. The UL version meets the US ultralight specifications, while the amateur-built version is heavier and faster.

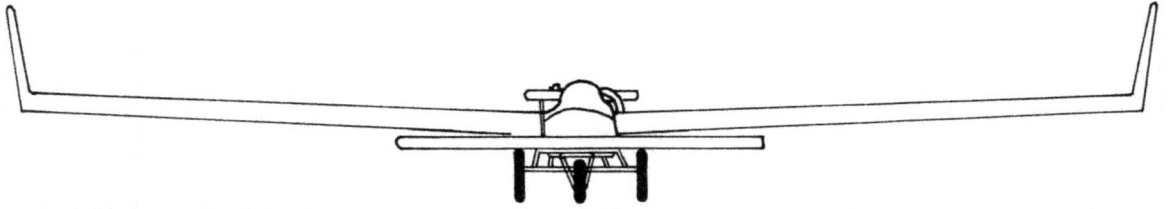

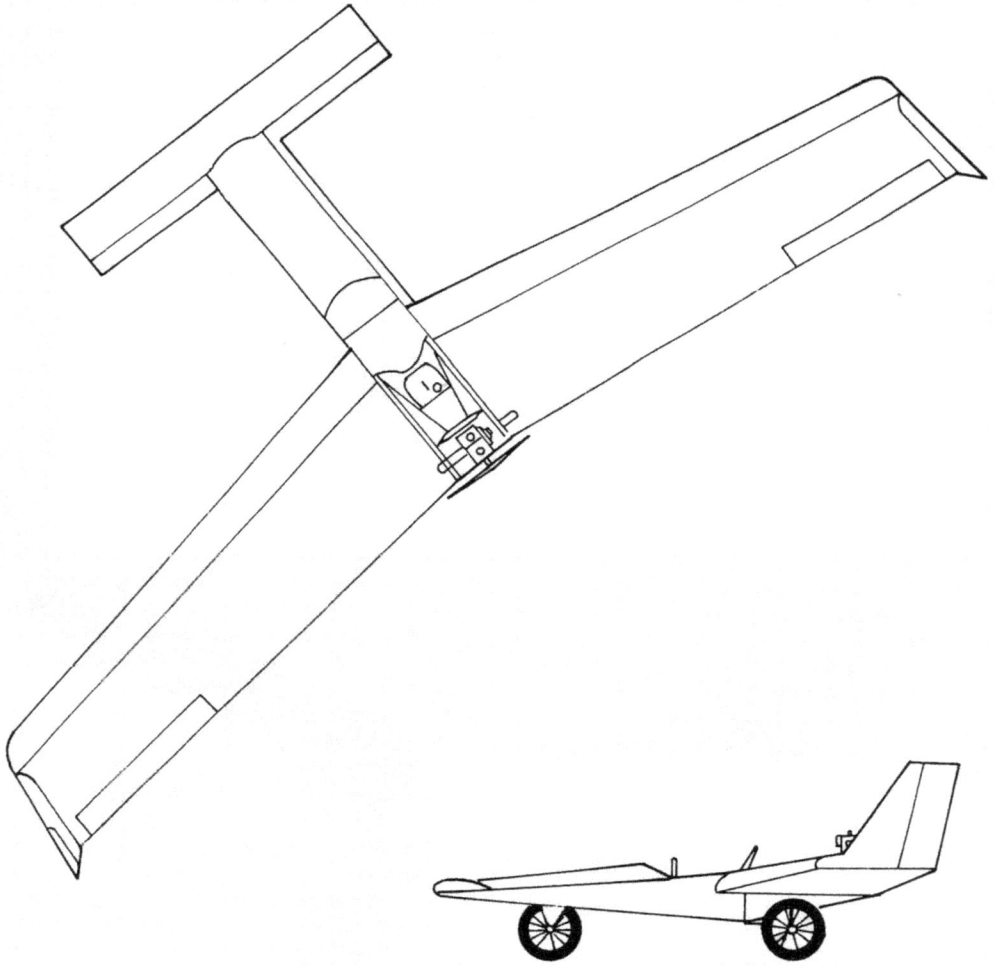

Fig. 3-35. Three view drawing of the Goldwing.

Fig. 3-36. The Diehl Aeronautical XTC is a very light amphibian of the ARV class. It is built entirely of composites, and offers good performance from the 25 hp KFM, horizontally opposed, two-cycle engine.

alternative to conventional spraying for the farm industry. A swath of 40 feet in width is covered at 60 mph.

Diehl XTC

The Diehl Aero-Nautical Co. of Jenks, Oklahoma designed and manufactures kits for their innovative single-seat XTC amphibians.

This straight-forward aircraft is 32 feet in span, 15 feet long, and has an empty weight of 304 pounds. FAA regulations now allow 20 extra pounds for outrigger floats and 30 extra pounds for hull floatability, so it still qualifies as an ultralight.

Power is a two cylinder opposed engine of 25 to 30 horsepower, pusher-mounted above the wing. The engine and its mount are fully exposed.

The hull has a shallow "V" bottom beneath the forebody, which ends in a deep step located about 25% of the aft wing chord from its leading edge. The afterbody sweeps up sharply to permit rotation for takeoff and landing. Hull beam is 31 inches, which yields a beam loading of 16.25 pounds per inch, at a gross weight of 550 pounds.

The landing gear is tricycle, with the main wheel legs rotating backward, partially enclosing the wheels in the wing for water landings. The nose wheel swings forward beneath the bow, and acts as a bumper. Forebody chines are equipped with spray strips.

The canard is mounted above the hull, almost in line with the pilot forward vision, but well away from the water. In the three view drawing, the canard seems all moving for pitch control, yet in the photo a slotted flap shows clearly at the canard tip.

143

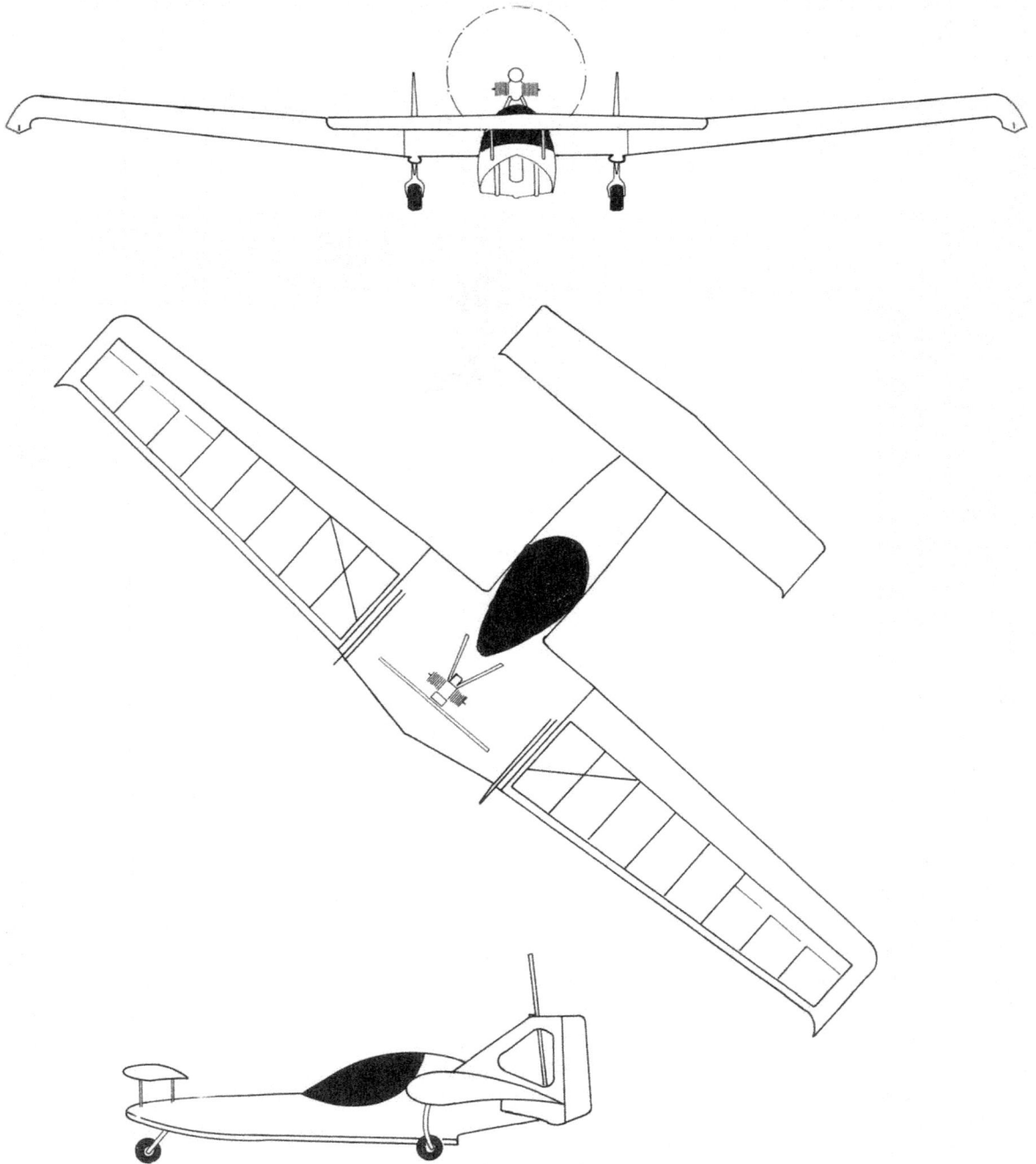

Fig. 3-37. Three view drawing of the Diehl Aeronautical XTC.

The wing is tapered and covered in clear plastic rearward of the "D" spar loading edge and outboard of the center section. The outboard panels are removeable for road transport. Wing tip (outrigger) floats provide lateral stability on the water and are cleverly shaped to function as vortex retarders in flight. Lateral (roll) control is through top surface spoilers.

Large fins, at the outboard ends of the wing center section, extend below the wings. The full length rudders act on both air and water for directional control in both mediums.

The cockpit is fully canopied in a very neat enclosure. Visibility is excellent, except for the canard. Construction is composite, and the aircraft offers STOL performance.

The wing area of 148 square feet yields a wing loading of 3.7 pounds per square foot at the gross weight of 550 pounds. Stall speed is a modest 25 mph, approach speed 30 mph, maximum speed in level flight is 61 mph, while never exceed speed is 80 mph. Cruise speed ranges from 35 to 55 mph. Rate of climb is 600 fpm and range, with 5 gallons of fuel, is 150 miles. With its electric starter, this plane would make an ardent fisherman's dream come true.

The XTC won 2nc place in the light aircraft category of the 1983 ARV design competition.

AMERICAN AIRCRAFT-FALCON

The Falcon is a strut-braced, high wing monoplane canard with a completely faired, open cockpit fuselage. It features a tricycle landing gear with a steerable nosewheel and three-axis aerodynamic controls including canard, elevator, tip rudders, and full span ailerons. Its high glide ratio allows engine-off soaring. It is available only as a completely ready-to-fly air vehicle.

The airframe makes use of advanced technology materials. The fuselage owes its light

Fig. 3-38. The American Aircraft Falcon is a fully enclosed, composite and aluminum construction canard, designed to meet the US ultralight specifications. It cruises at 60 mph and its clean lines afford it a glide ratio of 16-to-1, which should enable it to soar.

145

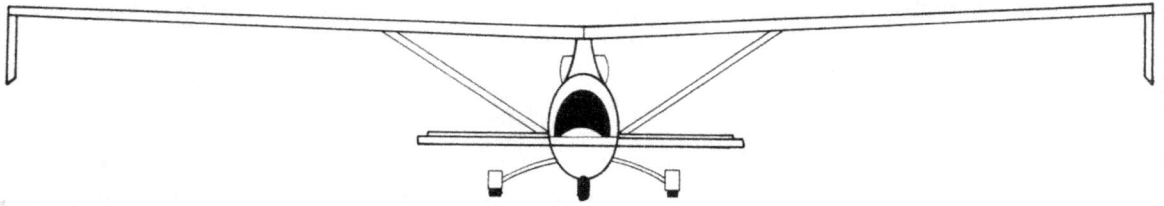

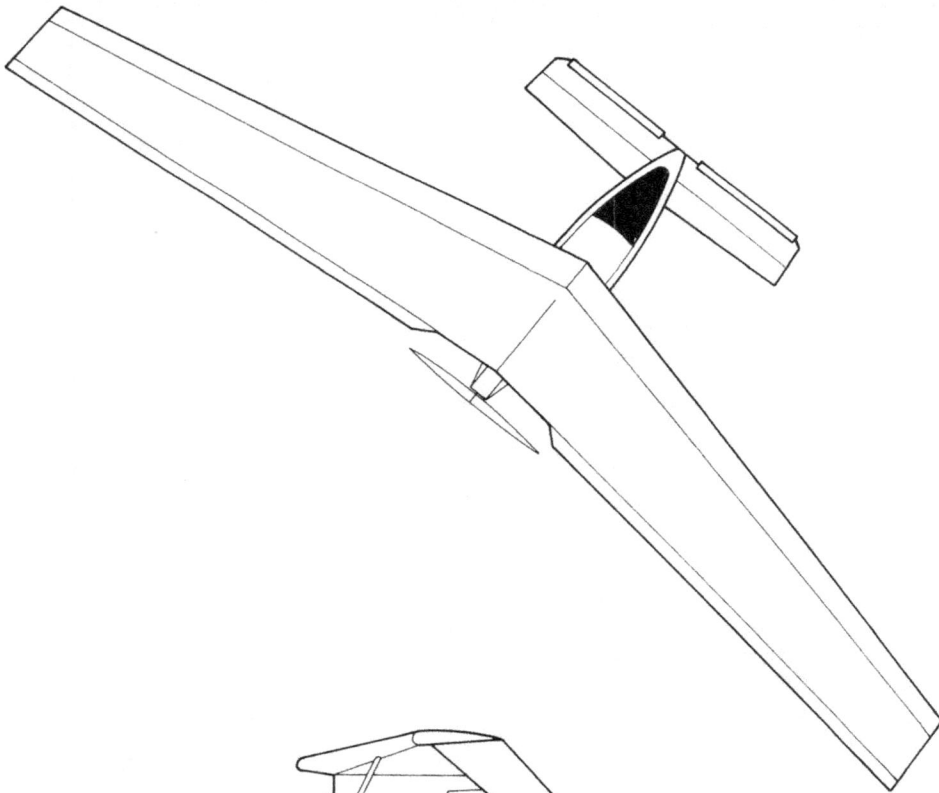

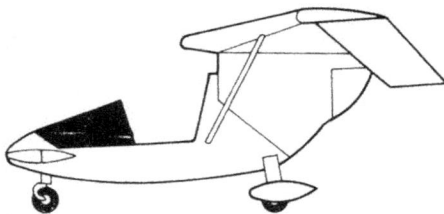

Fig. 3-39. Three view drawing of American Aircraft Falcon.

weight and strength to composites of kevlar, graphite and epoxy. The strut-braced, laminar flow wings consist of a D-cell leading edge, foam composite ribs, and an antidrag diagonal of aluminum tubing. The wing is covered with "Tedlar," a space-age alloy of teflon and mylar. It is nearly impervious to the elements, and affords the pilot with improved upward visibility.

The Falcon is powered by the single cylinder Rotax of 27 horsepower, which still enables the craft to cruise at 60 mph at a 40 mpg economy. The ship stalls at 26 mph and climbs at 600 fpm at sea level, with takeoff and landing requiring less than 300 feet. The crosswind capability is reportedly good. Field assembly/disassembly requires 15 minutes and the craft can be transported by trailer.

The air vehicle comes complete with the electronic Aerogage instrument panel, cabin heat, steerable nosewheel and brake, in-flight restarting, cockpit-operated choke, padded seat, and shoulder harness and seatbelt. The fuselage is available in four colors.

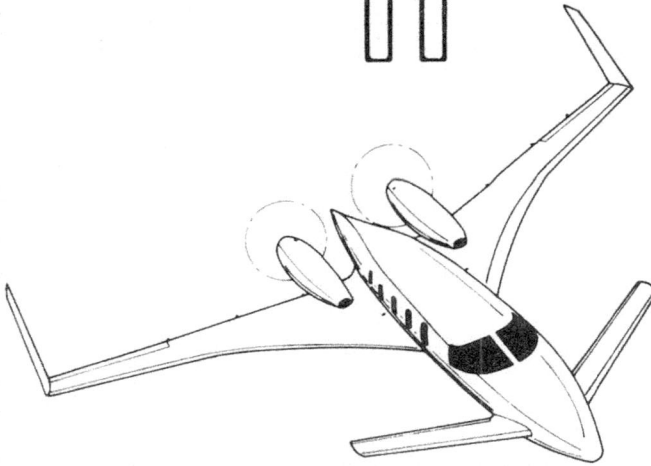

Sailplane And Motorglider Canards

This includes two aircraft, one a Turan design and the other a Swiss design by Avia Fiber of Switzerland. Both are canards, and both are aerodynamically very clean.

RUTAN MODEL NO. 77
SOLITAIRE MOTORIZED SAILPLANE

The Solitaire is a homebuilt, self-launching sailplane of canard configuration that won the Soaring Society of America's Self-Launching Competition in November 1980.

Since the greater part of a sailplane's life is spent utilizing only gravity for motive power, the wing design and airfoil sections are critical to its performance. John Roncz, a leading low Reynolds number airfoil designer, developed the airfoils for the Solitaire. Rutan selected Roncz 1052-177 and 1052-155 for the canard, which operates at Re of only 370,000 at low speed. For the aft plane, the Roncz 517-177 is used to eight feet from the center line, with the 515-140 outboard to the wing tip.

Because the main plane operates in the down wash and tip vortex upwash of the canard, a reduction in incidence of 2° takes place in the main wing from just outboard of the canard span to the wingtip. This provides an optimum span-wise lift distribution for the aftplane flying, as it does, in air already influenced by the canard.

Glide path control on a canard presents a problem. Flaps or spoilers on the aftplane would impact on its lift; the resulting pitch changes could be beyond the control capacity of the canard to correct. In typical Rutan fashion, the problem was overcome by a "spoil-flap" which is a one piece unit that acts as a flap below the wing trailing edge and as a spoiler above. When deployed, no pitch changing lift variations occur, but the added drag provides the glide path control desired.

The canard is mounted on the forward end of the fuselage for a minimum drag juncture. Similarly, the fuselage at the wing root, has no taper in plan view, so that little or no filleting is needed to avoid drag producing separation at the wing-fuselage joint.

Fig.3-40. The Rutan Solitaire is an amateur-built, self-launching sailplane. It won the Soaring Society of America's self-launching sailplane design competition. It features all composite construction, and a retractable, two-cycle engine. It dispenses with the typical need for a tow to altitude, and offers a glide ratio of 32-to1.

The pilot is located right on the CG—in the semi-reclining position his "belly-button" identifies the CG. Two wheels, bicycle fashion, partially enclosed in the fuselage, serve as landing gear—one behind the pilot and the other just under his knees.

The vertical fin and rudder are carried on a fuselage extension with ample clearance for rotation of the whole aircraft as it takes off or lands. Pilot vision is perfect. The aft wheel has a single brake. Small wing tip wheels mounted on tubular booms to the rear of the wing tips, provide lateral protection.

The canard is rectangular in plan, except at the tips and has trailing edge flaps for pitch authority. It is dihedraled for both stability and ground clearance, as the aircraft tips until the outboard wheels make contact, and the wing flexes slightly.

The aft wings are rectangular for roughly one third of their semi-span, then taper outboard to the tips. The spoil-flaps occupy the major portion of the rectangular span, and ailerons are located at the inboard end of the tapered portions.

The Soaring Society of America contest rules required static load testing, as proof of structural integrity, to 7.0-g's. All components survived this severe test without failure; the wing tip flexed a full 45½ inches, but returned to its normal relaxed position as the weights were removed.

The retractable power unit was a problem. Finally, a KFM 107E, two cylinder engine of 23 horsepower at 6,000 rpm with cylinders ar-

Fig. 3-41. The Solitaire in the powered mode. The prop stops in the vertical position, for stowage in the forward fuselage, along with the engine.

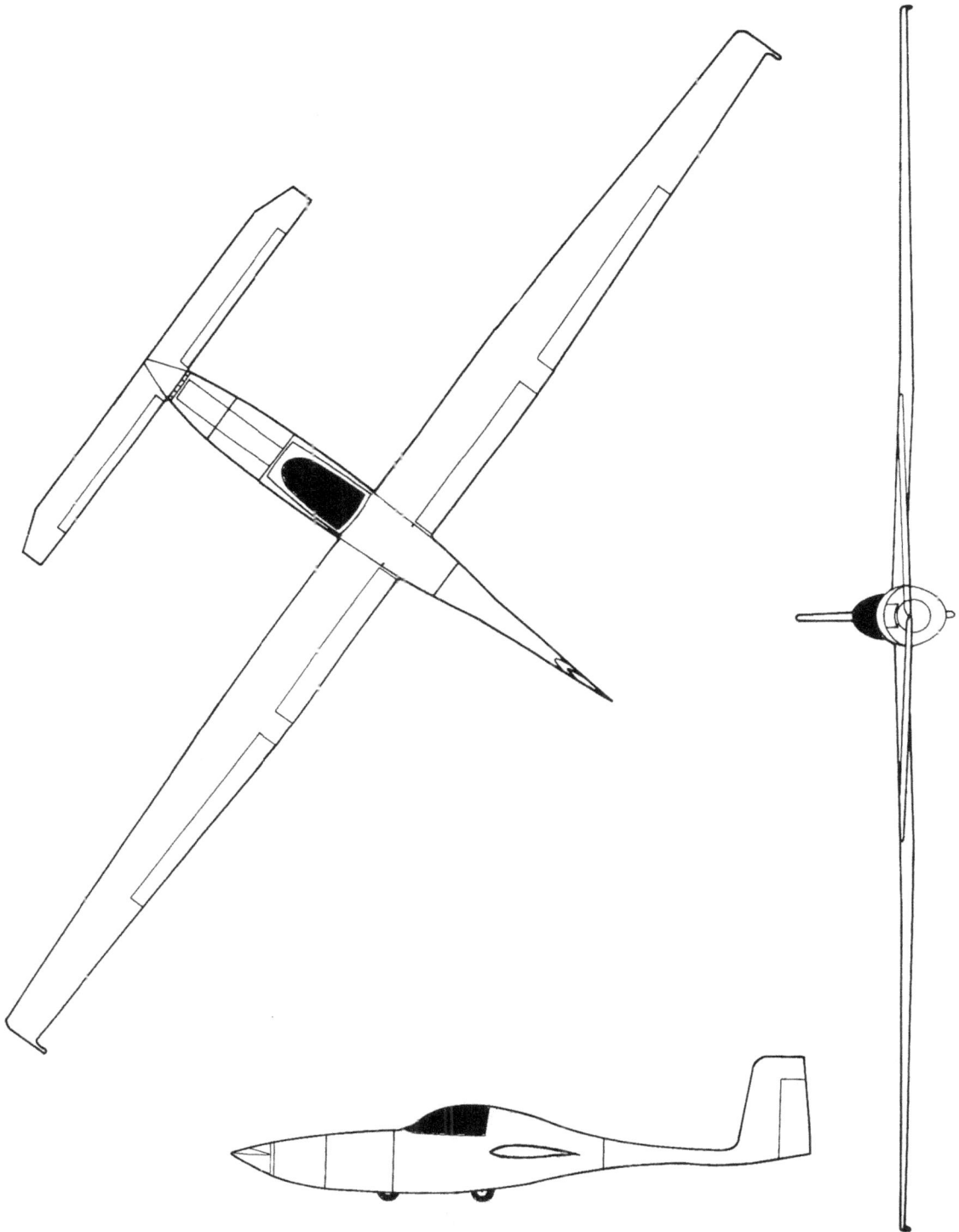

Fig. 3-42. Three view drawing of the Rutan Solitaire

ranged vertically rather than horizontally, was selected. An exhaust system was developed that would permit the engine, propeller, and exhaust to be retracted by pivoting forward into the nose. A cam that could be manually engaged with a lobe on the prop hub was incorporated so that the prop would windmill to vertical for retractions. The engine compartment doors are held closed with bungee cords. The engine and its support pushes them open as they lift and hold the doors open while the engine is running. The open doors added destabilizing vertical surface ahead of the CG and necessitated additional vertical tail height for suitable operation.

The retraction unit is electro-hydraulic, with double-acting ram, borrowed from outboard motor speedboat technology. This unit's travel is controlled by limit switches at the engine's upper and lower travel limits. Micro switches effectively disengage both starter and ignition unless the engine is fully up and in correct position to run. The engine can thus be raised, started or stopped, and lowered into the fuselage at will—in flight, or on the ground. The aircraft is independent of tow-planes, and can take off and fly or land and maneuver on the ground under its own power, or soar unpowered at the will of the pilot.

The solitaire has a span of 41.75 feet, and gross and empty weights of 620 and 380 pounds respectively. It has a wing area of 102.44 square feet and a wing loading of 6.05 pounds per square foot. Minimum flying speed is 37 mph, Vne 133 mph, and it carries five gallons of fuel.

This machine has all the solid, safe, stable, no stall/spin characteristics of its "stable-mates." The spoilflap descent control system is excellent. With engine folded, a glide ratio of 32-to-1 is achieved, providing true soaring capability.

The SSA presented Rutan its well-deserved "Outstanding Achievement" Award for the design of the Solitaire. Construction follows the Rutan developed composite materials and procedures as used on his VariEze, Long-EZ, De-Fiant, etc. Plans and kits are available.

Avia Fiber 2 FL.

This unique canard design was developed in Switzerland by Avia Fiber Aircraft Engineering and Manufacturing, Wald, Switzerland, headed by Hans U. Farner, formerly a Lockheed, California structural methods engineer. Structurally it follows European practice; Kevlar, Fiberglass, epoxy and foam sandwiches are produced in female production molds for wings and fuselage.

The Swiss Airworthiness Certificate labels the 2FL as an ultralight sailplane, proofed to regular sailplane standards—but without provision for powered plane aero towing. It is designed for foot-launching, or for a rolling takeoff on an optional wheel gear, down a slope of 1:5 pitch. A rolling takeoff requires 120 feet.

In a foot-launch, the pilot's legs project through an opening in the fuselage while the open canopy permits upright posture. At flying speed, the pilot leaps abroad to a prone position, then closes the leg doors and canopy.

Roll control is through "external airfoil" ailerons that appear more like the slotted Frize type. Yaw control consists of banking the canard wing plus or minus 5° relative to the main wing. Pitch or "speed" control requires the pilot to move forward or backward. Total movement is 32 inches, on a track-guided board. A speed range of 23 mph to 62 mph is obtained. Main wings and canard are fixed, incidence-wise.

The aftplane is supported on a V—"pylons," similar to the Waspair Tomcat. The trailing edges of the V are double split flaps which can be opened to provide drag for descent control without adding any pitch disturbing lift. The "V" surfaces also lift. Airfoils are Wortman FX 63-137 sections.

Landings are made on a retractable front skid. The pylon trailing edge airbrakes permit steep, slow approaches, similar to a Rogallo hang glider. The "V" pylons also provide directional stability.

The aircraft can be dismantled to a storing length of 15.8 feet. Built-in self-locking elements assure alignment and the control junctions are automatic.

Dimensions and weights are 44.4 feet span, 140 square feet wing area, and empty weight of 108 pounds, a payload of 110-265 pounds and, a wing loading of 1.47 to 2.54 pounds per square

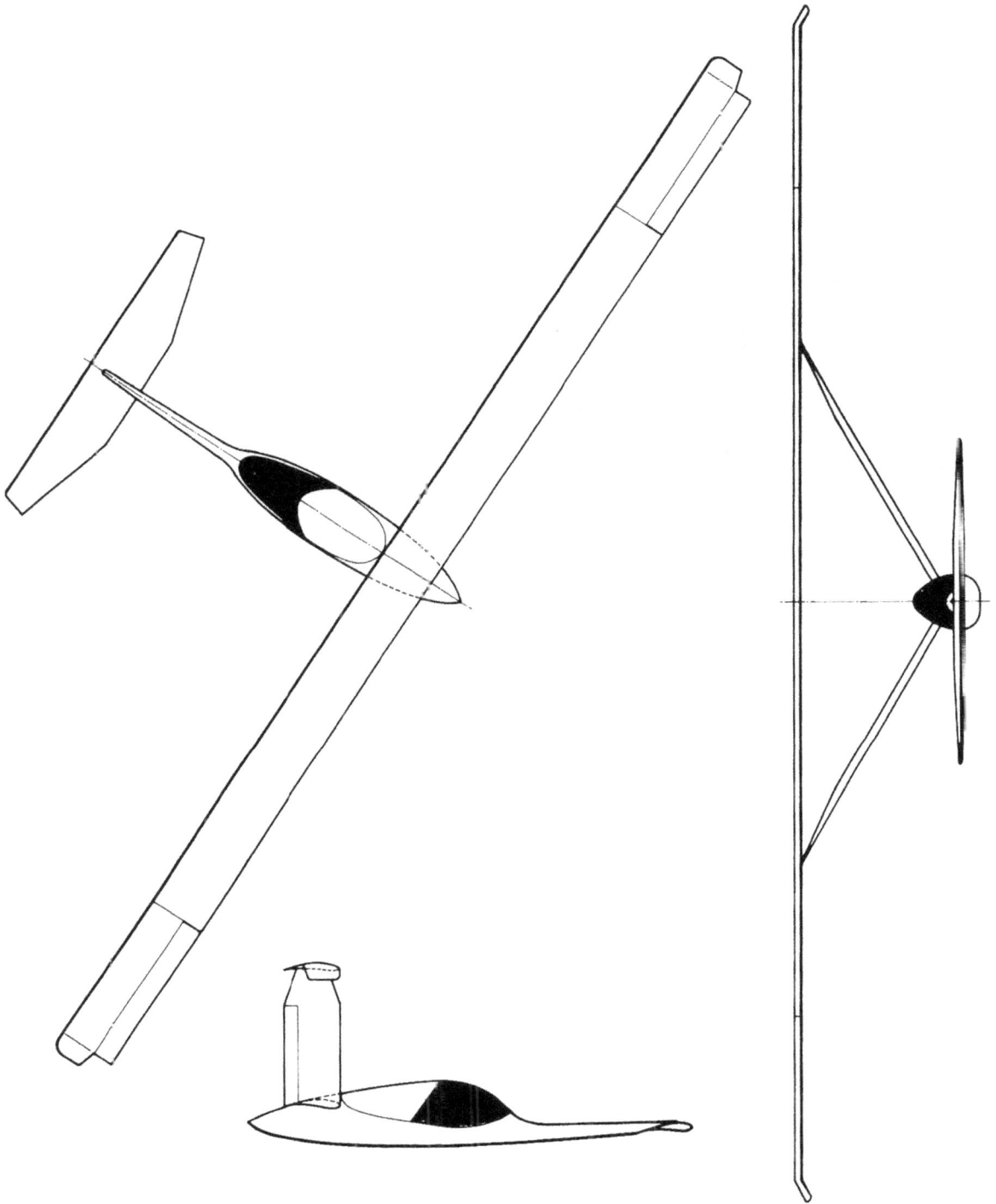

Fig. 3-43. Three view drawing of the Aviafiber Canard 2FL.

153

Fig. 3-44. The Aviafiber Canard 2FL is an extremely efficient, foot-launched sailplane. It is constructed of composite materials, following the European practice of using female molds for production. It can also be rolled down a slope for takeoff.

foot. Calculated performance figures are: a lift drag ratio of 31-to-1 at 35 mph, a minimum speed of 22.4 mph, a foot-launched speed of 12.4 mph, a minimum sink ratio of 94.7 feet/min. at 30 mph, and a maximum sink rate of 394 feet/min. at 62 mph.

The Canard 2FL has all the virtues of a well designed canard. It is stable, and free from un-controlled departures, stalls and stall/spins. The combination of an extremely narrow circling with low sink rate, is claimed to outperform conventional sailplanes, and equals them in straight flight sink rate. The neutral pitch behavior of the pylon speed brakes allows slow steep approaches for safe emergency landings in cross country flights.

12

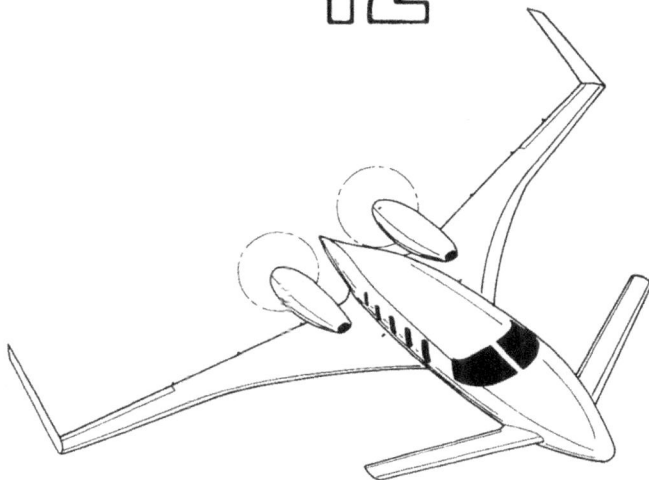

Human And Solar Powered Canards

In November 1959, Henry Kremer, an English Industrialist and devotee of physical fitness, offered a cash prize of 3,000 pounds for the first British aircraft to fly under human power alone. In 1967, the award was doubled, and opened to international contestants. In 1973, it was increased to 50,000 pounds, the largest aviation prize in history. The Kremer Prize rules required a sustained figure eight flight over a closed course, around two pylons one-half mile apart, with a ten-foot minimum altitude at the beginning and end of the course. The Royal Aeronautical Society specified the regulations and conditions of the competition, which was conducted by the Aircraft Owners and Pilots Association.

In the years from 1961 to 1976, fifteen aircraft were developed by individuals or groups from six countries: Australia, England, France, Italy, Japan and the U.S.A. The prize was finally won by the "Gossamer Condor," the U.S.A. entry designed by Paul MacCready and piloted and powered by Bryan Allen, a bicyclist and athlete

six feet in height and 137 pounds in weight.

It is intersting to note that MacCready's Gossamer Condor had the largest total wing area, the lowest empty weight and at gross weight, the lowest wing loading of any of the most nearly successful competitors on record. The figures were: wing area 835 square feet (including canard area), 96 feet in span, empty weight 70 pounds; a flying weight of 207 pounds and a wing loading, at flying weight, of 0.25 pounds per square foot (or 4.0 ounces/square foot—lower than most model airplane gliders).

After MacCready's win, Kremer announced another prize, this one for 100,000 pounds, for the first contestants to successfully fly the English Channel from England to France.

The rules announced by the Royal Aero Club called for a takeoff from anywhere in England, provided its height did not exceed 50 meters above the Channel surface. The obvious departure point would be Dover, with a landing at CAP GRIZ-NEZ, 22 miles away and something over two hours of flying in still air, at an airspeed of around 10 mph.

The prize was again won by MacCready on June 12, 1979 with his "Gossamer Albatross," piloted and powered by Bryan Allen who almost "ran-out-of-steam," but succeeded in landing on the beach of France at the Cap.

THE GOSSAMER CONDOR

This MPA (man powered aircraft) has its flying surfaces and streamlined pilot enclosure covered in Mylar, ½ mil in thickness, except on the bottom of the wing where ¼ mil thickness is used.

All aluminum tubing is 6061-T6; the wing spar being built of 12 foot sections of 2 inch outside diameter tubing. The 0.020 inch wall thickness at the wing root diminishes to a .015 inch wall at the wing tips.

Main wing ribs are built up from ¼ inch OD X .020 inch wall tubing. The canard ribs are

spruce and balsa. The wing leading edge is corrugated cardboard, while the canard is styrofoam sheet expanded and formed in an oven.

The propeller was built from balsa ribs, leading and trailing edges, and covered in Monokote. The prop spar was one inch OD X 0.20 wall tube. It turned at about 120 rpm, geared up from a pedal speed of 90 rpm. Airspeed measuring equipment consisted of an Erisman pacemeter adapted to a propeller mounted on the boom connecting wings and canard.

In all, 70 exterior wires support the wing and canard, running from the top of an 18 foot kingpost, and from the bottom of the pilot enclosure.

Aerodynamically, the aft streamlined pilot housing provides directional stability. The wings, with their gentle sweepback, twist, and reflexed wing section, computor designed by Dr. Peter Lissaman, were relatively stable. The canard forward surface provided pitch, as well as direc-

Fig. 3-45. The MacCready Gossamer Condor was the first man-powered aircraft to fly the Kremer figure-eight course, and won $86,000 in so doing. The 70 pound craft covered the course in under 6½ minutes, at an average speed of 10.82 mph. The aircraft now resides at the National Air and Space Museum in Washington, D.C.

156

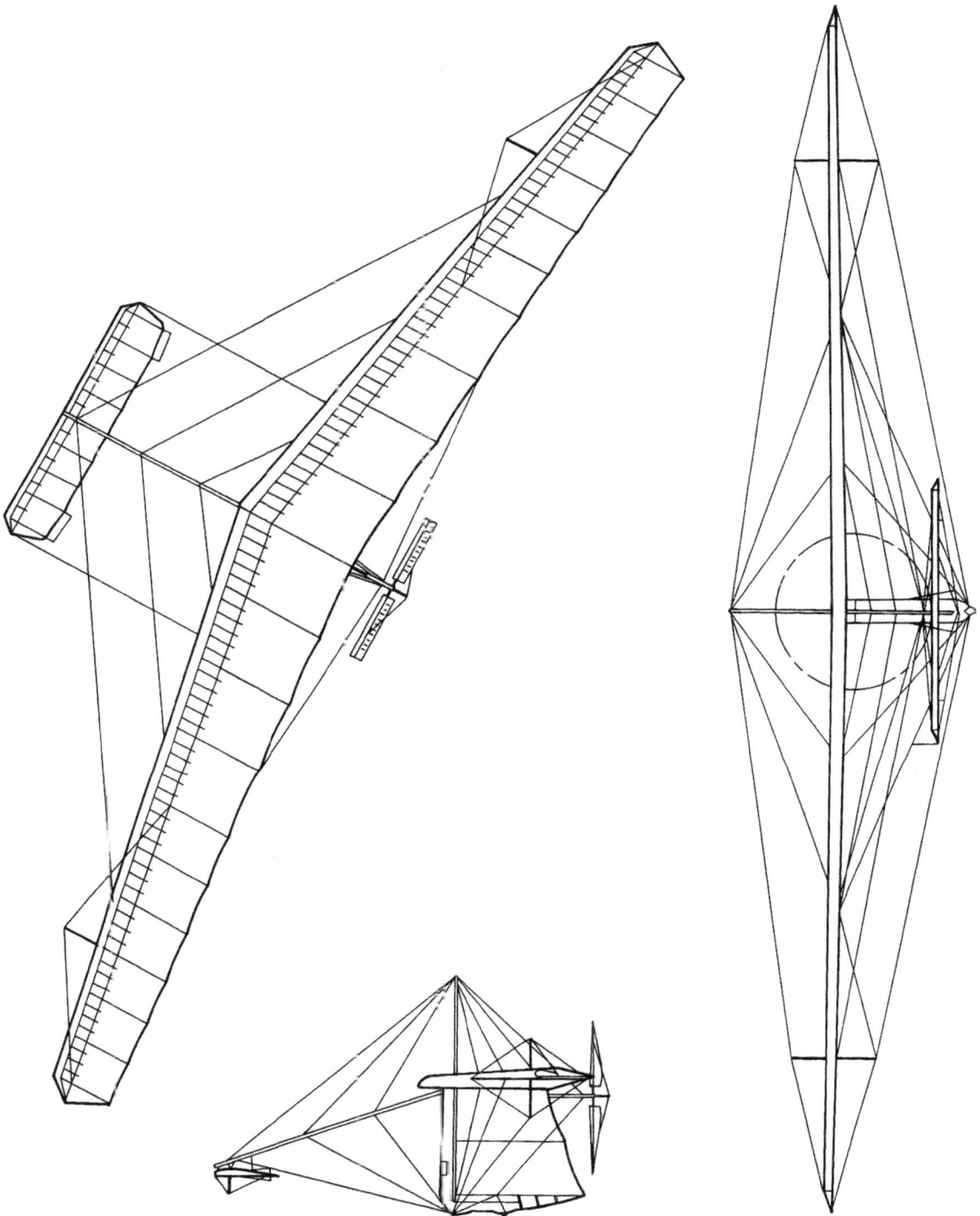

Fig. 3-46. Three view drawing of the Gossamer Condor.

tional control. It could be rolled by small ailerons to provide a lateral lift component, for small flight path corrections. For large turns, wing warping is employed, but due to the large span and slow speed, the warp is opposite normal. The tip on the inside of the turn is increased in incidence to add drag and to compensate for its slower air speed. The added drag causes a yaw and the yaw causes roll. Turns are very gentle, the angle of bank being 3°, for a turning radius of roughly 250 feet. Steeper banking leads to skidding.

The canard is set at a lower angle of attack than the main wing to reduce trim drag. Therefore, the wing stalls first, without warning, at 9.5 mph. Needless to say, this would be a fatal trait in a normal canard airplane.

Hard pedaling and stick forward results in a slow recovery, accompanied by an altitude loss of 10 feet. A more vigorous stall results in a para-chute-like descent, which can be damaging to the plane, but seldom resulting in pilot injury.

The airplane called for a high expenditure of human effort to maintain flight. The maximum Bryan Allen could pedal it was 18 minutes before he was exhausted. The pedals are connected to the propeller by a plastic drive chain running over plastic pulleys.

During its development, many incidents caused damage to the frail structure, but repairs were quickly accomplished. Flying could only be done in "no wind" conditions.

The Condor is now on display at the National Air and Space Museum in Washington, D.C., close to the Wright brothers' Kitty Hawk Flyer.

GOSSAMER ALBATROSS

The Albatross was humorously called the "Plastic Pigeon" and the "DuPont Dove"

Fig. 3-47. The Gossamer Albatross was a much refined evolution of the Condor. It was built in response to Kremer's announcing a prize of $200,000 for the first man-powered aircraft to fly the English Channel. The Albatross was much improved over the Condor, employing a more efficient structure and improved aerodynamics. It required only ¼ hp to remain airborne. Photo shows it being test flown at Shafter, California. NASA photo.

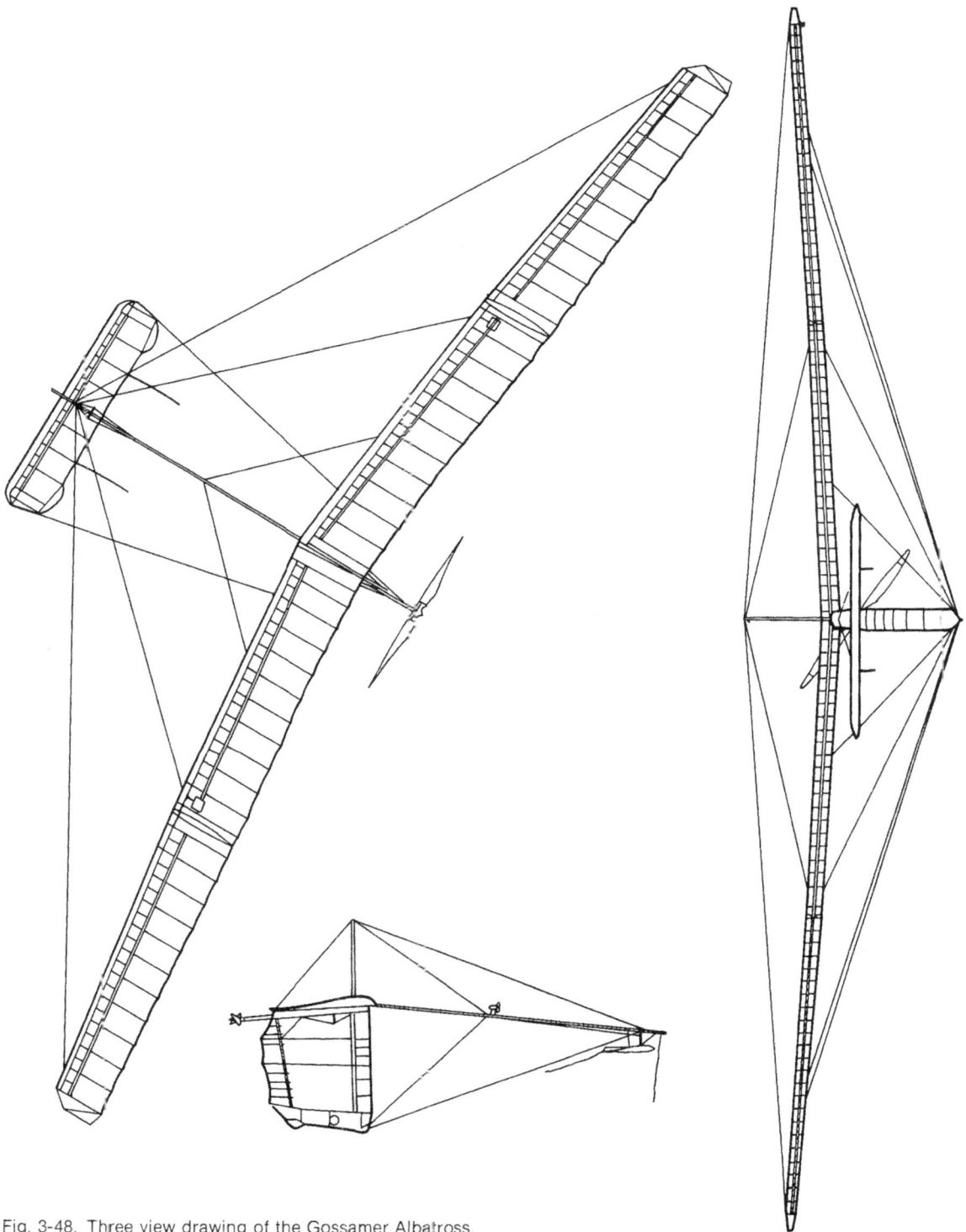

Fig. 3-48. Three view drawing of the Gossamer Albatross.

because of the many components of DuPont manufacture used in its construction. These included: "Mylar" covering; Polyaramid fabric reinforcement, "Kevlar" chord control cables running over "Delrin" pulleys; Polyurethane drive chain and structural members made up of ⅜ inch wide strips of 0.013 inch thick graphite composite bonded to ¼ inch thick Polystyrene foam.

DuPont Corporation also provided $200,000 to defray the costs involved in the development, testing, manufacture, travel and living expenses preparatory to the cross-channel success. This also made possible two back-up Albatross's, a working base in Britain, communications and escort vessels, etc. and transfomed a shoe-string hope into a well-organized, successful operation. DuPont clearly recognized the benefits to be de-rived from the world-wide exposure given its products in this venture.

The Condor's 7½ minute prize winning flight had required great physical effort—far too much for a 2 hour, 22 minute cross-channel flight. Dr. MacCready realized that 0.25 horsepower was the maximum human output possible for any length of time. To accomplish this remarkable feat, Mac-Cready reduced weight, while increasing the wing's stiffness with less bracing. He also went to a smaller area, higher aspect ratio wing, with a better airfoil.

To Bryan Allen must go the ultimate credit for the Albatross' successful channel crossing. Dehydrated, without instruments, flying in a steamed up capsule, and fighting increasing turbulence and head winds, he perservered for 2 hours and 49

Fig. 3-49. The Gossamer Albatross, piloted and powered by Bryan Allen skims over the surface of the English Channel. NASA photo.

Fig. 3-50. The Gossamer Penguin was a smaller, lighter version of the Albatross. It made the first man-carrying, solar-powered flight on May 18, 1980. Note solar panels aimed at the sun.

minutes. He arrived exhausted and almost incoherent, but successful, on the sandy beach of Cap Griz-Nez, to claim the prize.

GOSSAMER PENGUIN

The Penguin is a smaller version of the channel-crossing Albatross. It is considerably lighter, and was designed to be flown by a lighter pilot, in this case Paul MacCready's son Marshall.

Power was an Astro 40 model airplane electric motor, developed by Astro Flight. Electricity was provided by four solar panels with the total of 400 watts available from 3,640 solarcells. The 11 foot diameter propeller turned at 120 to 130 rpm with a speed reduction from the motor of 129 to 1.

Specifications of the Gossamer Penguin were: wing span of 71 feet - 1 inch, a wing area of 312 square feet, a propeller diameter of 11 feet x 14 feet pitch, an empty weight of 68 pounds, and a gross weight of 168 pounds.

The historic solar powered flight was made May 18, 1980, the first manned aircraft was flown on power generated from solar cells alone.

Section IV
Future Canard Developments

13

General Aviation Canards

General aviation is finally starting to awake to the aerodynamic and structural advantages as demonstrated by Burt Rutan's aircraft. Thanks to thousands of amateur builders, as well as to Rutan's SCALED, Inc. developments, particularly the Microlight Canard and Conventional Fairchild Trainer, they are beginning to take notice. Currently, four aircraft companies are developing canards (one is actually a hybrid). Furthermore, Rutan has also laid down a design for a commuter transport aircraft based on his canard concepts.

The following pages deal with the five new designs.

RUTAN COMMUTER MODEL 78-1

Model 78-1 was designed as a 36 passenger, twin engine airliner to provide safe, quick intercity shuttle service.

The unique features of this aircraft are the swept forward aft wing, with flaps and elevators, while the foreplane is also equipped with flaps and elevators. A center line thrust arrangement is accommodated by the turbo prop engines mounted on the fin—one is a pusher, the other a tractor. It reflects the Fowler flap development pioneered in the Grissly design.

The tricycle landing gear main wheels retract into streamline fairings at the lower aft fuselage sides while the nosewheel retracts into the aircraft's nose. Wing tip winglets are employed but the rudder is beneath the rear engine. The only conventional features are the ailerons.

The Model 78-1 spans 70 feet, is 65 feet long and burns 505 pounds of fuel per hour. Cruising speed is over 300 mph, climb is 2110 fpm on both engines, and 770 fpm on one. Landing speed is 94 mph.

A scale model is being wind tunnel tested at NASA's Langley Research Center.

It is suspected that Rutan's work on the Beechcraft Starship I has resulted in model 78-1 being "put on the back burner."

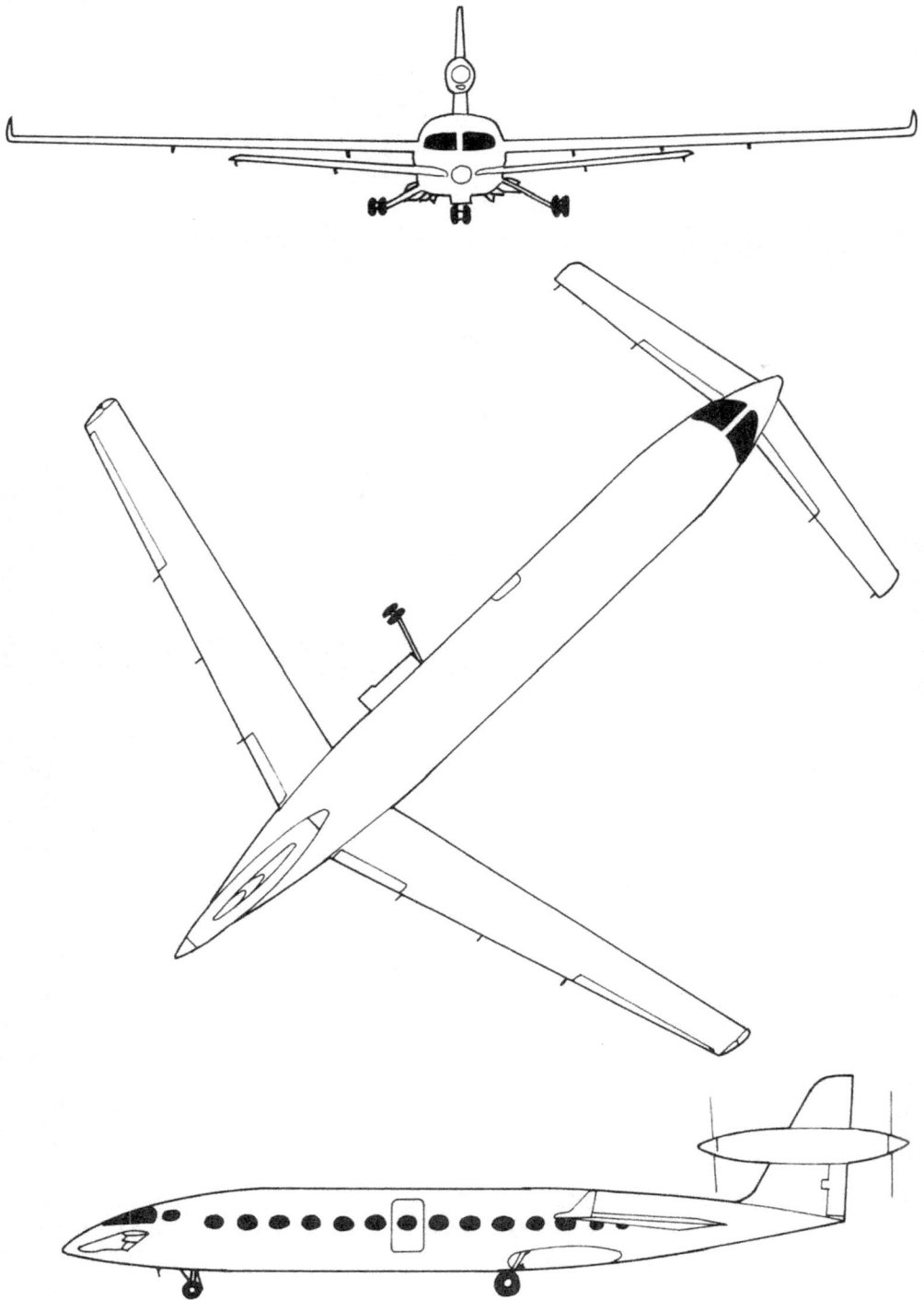

Fig. 4-1. Three view drawing of the Rutan Model 78-1 Commuter.

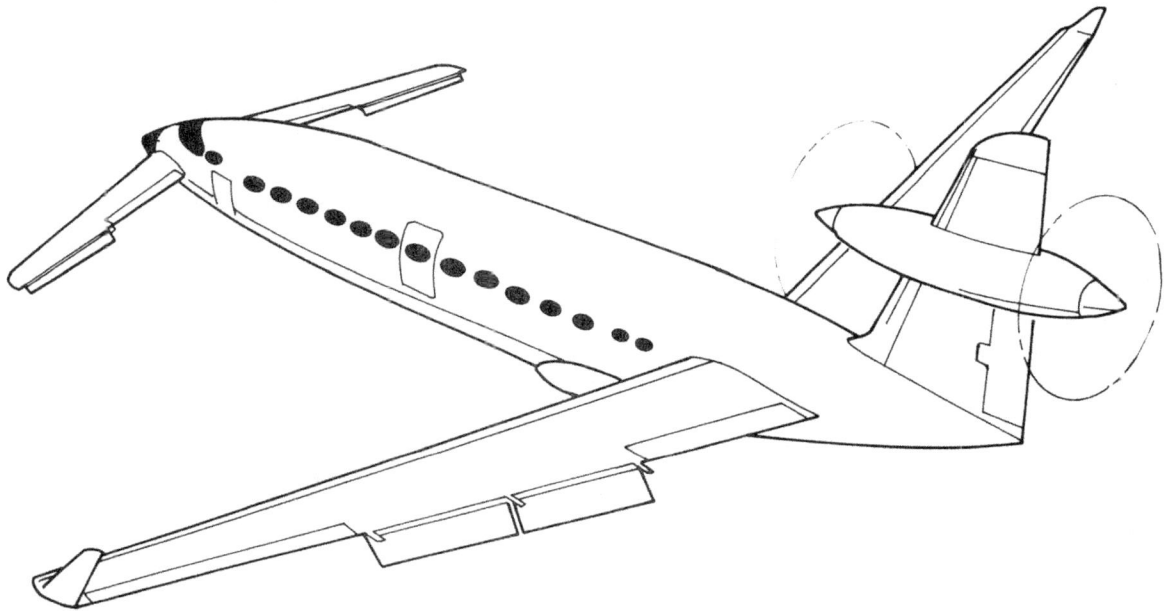

Fig. 4-2. Artist's rendition of the Rutan Model 78-1 Commuter. The forward sweptwing aircraft, was intended to accommodate 36 passengers.

BEECHCRAFT STARSHIP I
CORPORATE JET-FAN

Beech Aircraft Corp. is one of the oldest and most respected of the major U.S. general aviation manufacturing companies. While conservative in its developments, Beech has never shied away from the unconventional if there were sound reasons to do so. The pre-World War II "Stagger-wing" and the postwar "Bonanza," with its V-tail, are prime examples of that thinking. Nevertheless, adoption of a canard of advanced composite construction is a bold move that will challenge both the FAA and Beech engineers in its certification.

After initial studies of tandem wing pusher lay-outs, Beech invited Rutan to join in the study. Their joint efforts culminated in Rutan's SCALED Composites, Inc. construction of an 85% scale "proof-of-concept" flying prototype that began flight tests on August 29, 1983. At the National Business Aircraft Association's Convention in Dallas in October 1983, its flight demonstration

and, the showing of a full scale, non-flying mock-up, were the "hit" of the show.

Beech is now building six full scale airframes, including two for ground testing.

Beech's goals for the Starship 1 are:
- Higher operational speeds and altitudes than current business turboprop types.
- Improved fuel economy.
- Lower cabin noise levels.
- Stability in all flight regimes with no stall/spin.
- Less engine-out, off-center thrust problems.
- Pressurization of cabin, and more interior height and length.
- A docile, utilitarian and good field performance aircraft.

The Starship accommodates two pilots and nine passengers, a refreshment center, toilet, and two baggage compartments that are in-flight accessible.

It is to be powered by two Pratt and Whitney PT6A-60 turbo props flat rated to 1,000 shp each for takeoff, driving four bladed pusher propellers.

Fuel is accommodated in the blended aft "wet wing" on or close to the CG.

Dimensions are: span 54 feet-7 inches, length 45 feet-5 inches. It has a maximum takeoff weight of 12,500 pounds.

Performance figures include a cruising speed over 400 mph, initial rate of climb of 3,300 fpm, a ceiling of 41,000 feet, and a range of 2,070 miles with 45 minutes reserve.

The aft wing has two sets of Fowler flaps mounted on three external flap guides on each side. The swept back forward wing is interconnected to the flaps and sweeps forward when flaps are lowered, for slow, short landings.

The wing is very "Long-EZ" in plan, has large "tip-sails" (winglets) incorporating rudders for directional stability and control. The ailerons are "conventional" and airfoils are by John Roncz.

The two pusher engines are spaced close to the fuselage to reduce engine-out asymetrical thrust. No large central fin is employed, but there is a small ventral fin and rudder. The dual main wheels retract inward into the wings, and the nose wheel retracts forward into the fuselage nose.

The flight deck will also be advanced design, with a Collins "all glass" panel incorporating CRT's for flight, navigation and performance monitoring systems. Single pilot operation certification is planned, despite provision of two seats and full dual controls.

The 85% of scale prototype is undergoing extensive flight testing at Mojave, California to provide Beech engineers with the design's flight characteristics before the full scale prototype flies.

That Beech would enlist the aid of Rutan's Scaled Composites, Inc. is, in itself, recognition that Burt Rutan has achieved the status of a lead-

Fig. 4-3. Beechcraft startled the aviation world when they unveiled their incredibly beautiful Starship 1 in 1983. The Rutan designed original, features a variable sweep canard that works in concert with the main wing's Fowler flaps, producing a high lift system without trim changes or shifts in the Neutral Point. Pitch control is via canard elevator, supplemented by aftwing elevon effect from the ailerons.

Fig. 4-3a. The incredibly beautiful Starship looks like something out of the movie "Star Wars," and it is quite efficient.

Fig. 4-4. The Starship 1, as seen in these photos, is actually an 85% scale flying prototype. With it, engineers gathered pertinent data on performance, stability and control, enabling them to better design the full scale version. It will carry two pilots and nine passengers.

ing innovator and designer in world aviation—a well deserved accolade.

Beech management is recognized widely for the courage and foresight needed for this dramatic leap forward in configuration and technology. Beech Aircraft Corporation is a subsidiary of Raytheon Company.

AVTEK 400

The Avtek 400 could, humorously, be called the "DuPont Duck," being a canard (French for duck) and constructed of DuPont's Kevlar and Nomex composites, along with Dow Chemical resins.

It was first displayed in mock-up form at the 1983 Paris Air Show. Its composite structure permits an empty weight of only 3,100 pounds. Maximum takeoff weight is 5,500 pounds, or 77% of empty weight for payload.

The aircraft has a somewhat conventional appearance, is 34 feet in span, is 34 feet long, and is powered by two Pratt and Whitney 680 shp PT6A-28 turboprop pusher engines, mounted in nacelles supported by short pylons over the wings. The propellers are Hartzell three-bladed constant-speed, full-feathering, reversible pitch pushers with Q-tips.

Fuel is carried in integral fuel tanks, 100 gallons in each wing and 40 gallons in, oddly enough, the foreplane. Fuel management in flight is automatic, maintaining the CG in its optimum location. In addition to the pilot, the aircraft will accommodate five to eight passengers.

Estimated performance figures are: a maximum speed of 425 mph, maximum cruise speed of 415 mph at 25,000 feet altitude, an economy cruise speed of 300 mph at 33,600 feet altitude, a service ceiling of 38,000 feet, full load range of 2,600

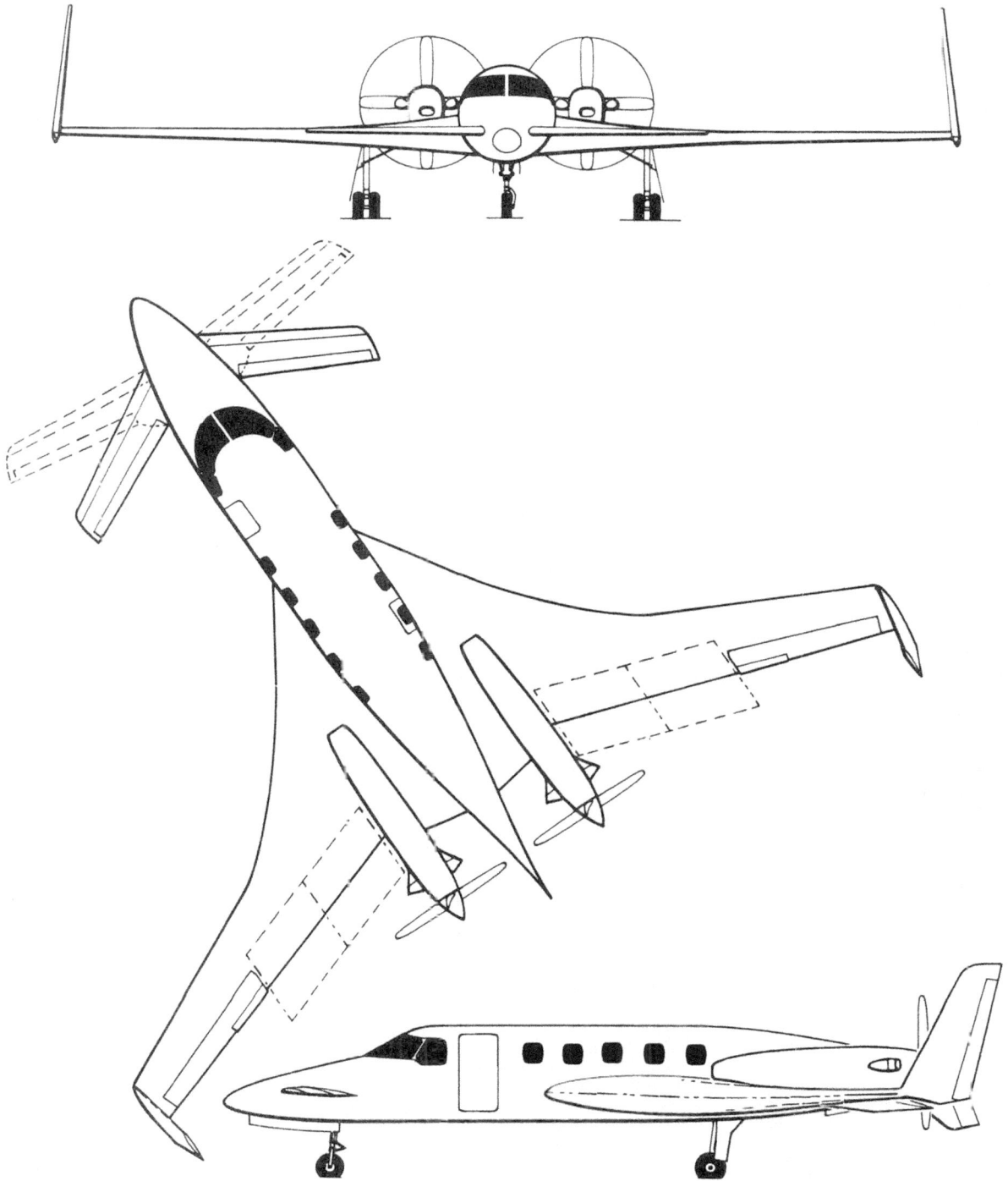

Fig. 4-5. Three view drawing of the Starship 1, 85% scale flying prototype.

Fig. 4-6. The Starship's flight deck includes an "all glass" panel with CRTs for flight, navigation, and performance monitoring.

statute miles, and a stall speed-power off of 89 mph. Climb with both engines is 5,226 fpm, and with one engine out, it is 2,370 fpm. Fuel consumption is 13.1 mpg.

The total wing area is 141 square feet of which the canard is 44 square feet or 31% of gross area. Wing loading is 29.8 pounds per square foot at gross and power loading is 3.12 pounds per shp.

The aircraft accommodations include a bar, lavatory, and generous baggage space. It offers unique facilities as an air ambulance, and for cargo carrying of up to 2,000 pounds in 140 cubic feet.

The aft wing has an aspect ratio of 8.2, 4° dihedral and is set at an incidence of 1°. Winglets are incorporated along with full span trailing edge controls in two sections of the wings. The outer sections serve as ailerons, while the inner sections are for "pitch stabilization." Trailing edge flaps are not used. Pneumatic de-icer boots are incorporated into the wing leading edge.

The foreplane is mounted on the upper fuselage just aft of the windscreen and is equipped with conventional elevators. De-icing is electric or engine bleed-air. The fuselage and baggage compartments are pressurized, the vertical surface is swept back and incorporates the rudder. De-icing of the fin leading edge is pneumatic. Landing gear is tricycle—the main wheels retract hydraulically inward into the wing while the nosewheel retracts forward and is steerable.

A wide range of optional avionics and color radar are offered. Full IFR instrumentation is optional.

Avtek is funded entirely by private investors. It was founded and is headed by Robert A. Dickes, a retired TWA Captain, and is designed by Al Mooney. Leo Windecker of "Windecker Eagle" fame, is VP of Research, and William W. Taylor IV, is VP of Operations. Assembly will be done by Aeronca, Inc. Pineville, North Carolina. Avtek Corp. is headquartered in Camarillo, California.

It is interesting to note that a number of radio controlled scale models of the Avtek 400 were built in various sizes, during the design's development. Certification is hoped for by the end of 1985.

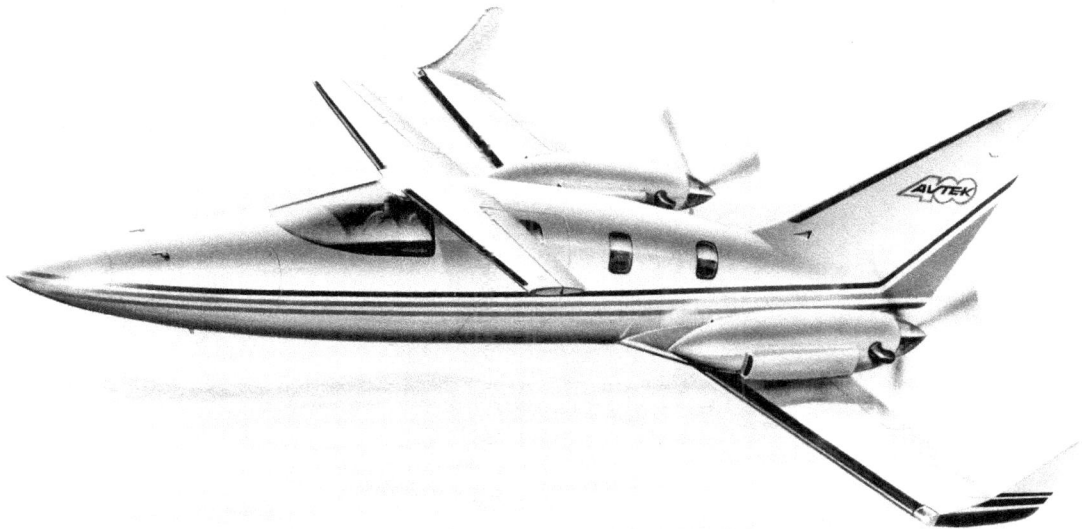

Fig. 4-7. The AVTEK 400 is another composite canard business aircraft. It will accommodate five to eight passengers. The location of its foreplane permits unrestricted forward and downward visibility.

Fig. 4-8. The Kevlar fuselage of the Avtek 400 featured accommodations such as a bar, lavatory, and generous luggage space. It will carry up to 2,000 pounds in 140 cubic feet.

Fig. 4-9. The Avtek's flight deck features a mixture of CRT, analog, and digital instruments. Full IFR instrumentation is available. The seats are composed of individually adjustable modules for improved circulation and long range comfort.

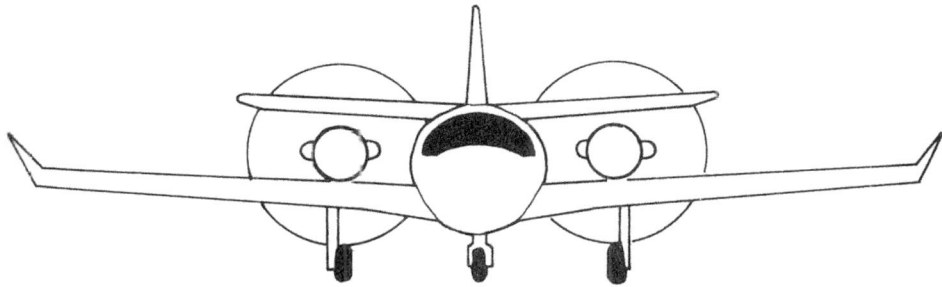

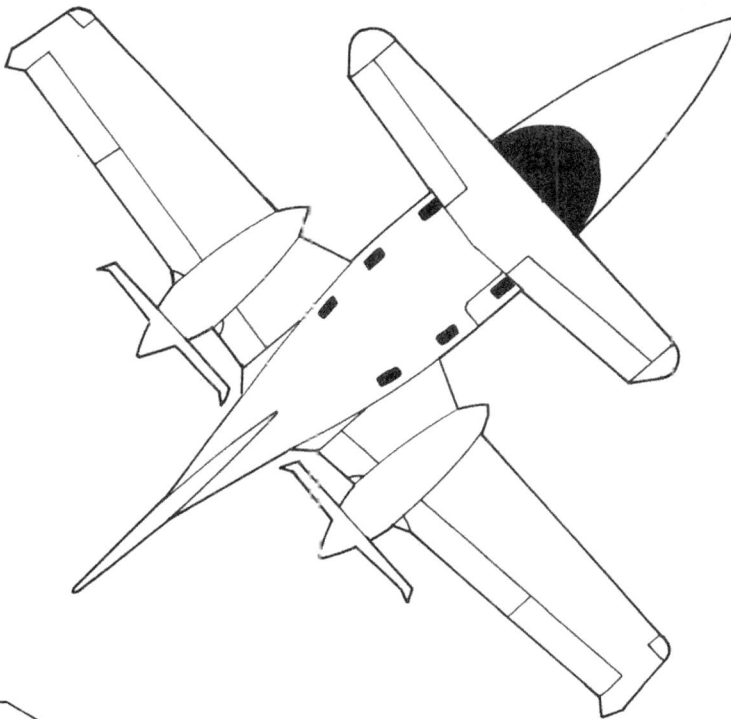

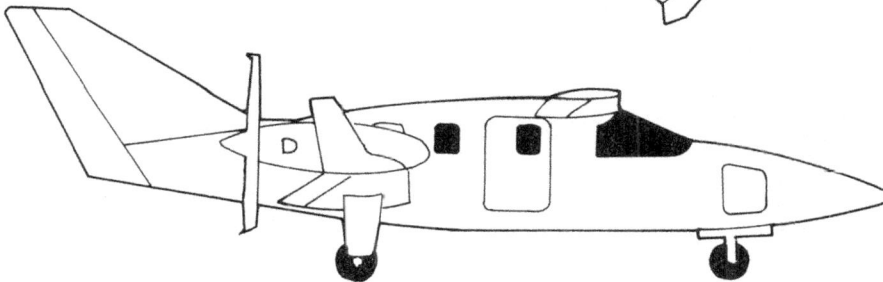

Fig. 4-10. Three view drawing of the Avtek 400.

OLD MAN'S AIRCRAFT CO.
"OMAC I"

First flown in December 1981, the Omac I is not only the oldest design, but also the only single engine canard. It is also the sole all metal airplane in this group of unusual General Aviation business aircraft.

The company name was inspired by the ages of Carl Parise and his original partner Larry Heuberger, now no longer associated with the company. Parise has a new partner in Steve Mihaylo, but is working board chairman.

The OMAC I's power unit is an Avco Lycoming LTPIOI-700A-I turboprop of 700 shp. The airplane is designed to carry, in comfort, six to eight persons, at a maximum cruise speed of 272 mph. Long range cruise speed is 230 mph, while maximum gross weight is 6,000 pounds. The power loading is 8.57 pounds per shp.

The empty weight of 3,200 pounds, leaves a 2,800 pounds useful load. A fuel capacity of 285 gallons provides a range of close to 3,000 miles with 45 minutes fuel reserve, and six people aboard. Ceiling will be 25,000 feet, pressurized to 8,000 feet, with a pressure differential of 5.5 pounds per square inch.

Wing span is 35 feet, total lifting surfaces, including the canard, are 294 square feet, while the wing loading overall is 20.4 pounds per square foot, the canard is loaded 30% more heavily than the main plane for stability. The propeller is a Hartzell three-bladed Q-tip, 90 inches in diameter. The aircraft is 29.5 feet in length. The landing gear is tricycle, with all three legs folding up into the fuselage.

The wing center section is more obviously Delta shaped than Rutan's designs. The constant chord outboard panels sweep back 10° and carry the

Fig. 4-11. The OMAC-1 was the first modern canard business aircraft to fly, back in 1981. It is also the only all metal and single engine powered of the new General Aviation canards.

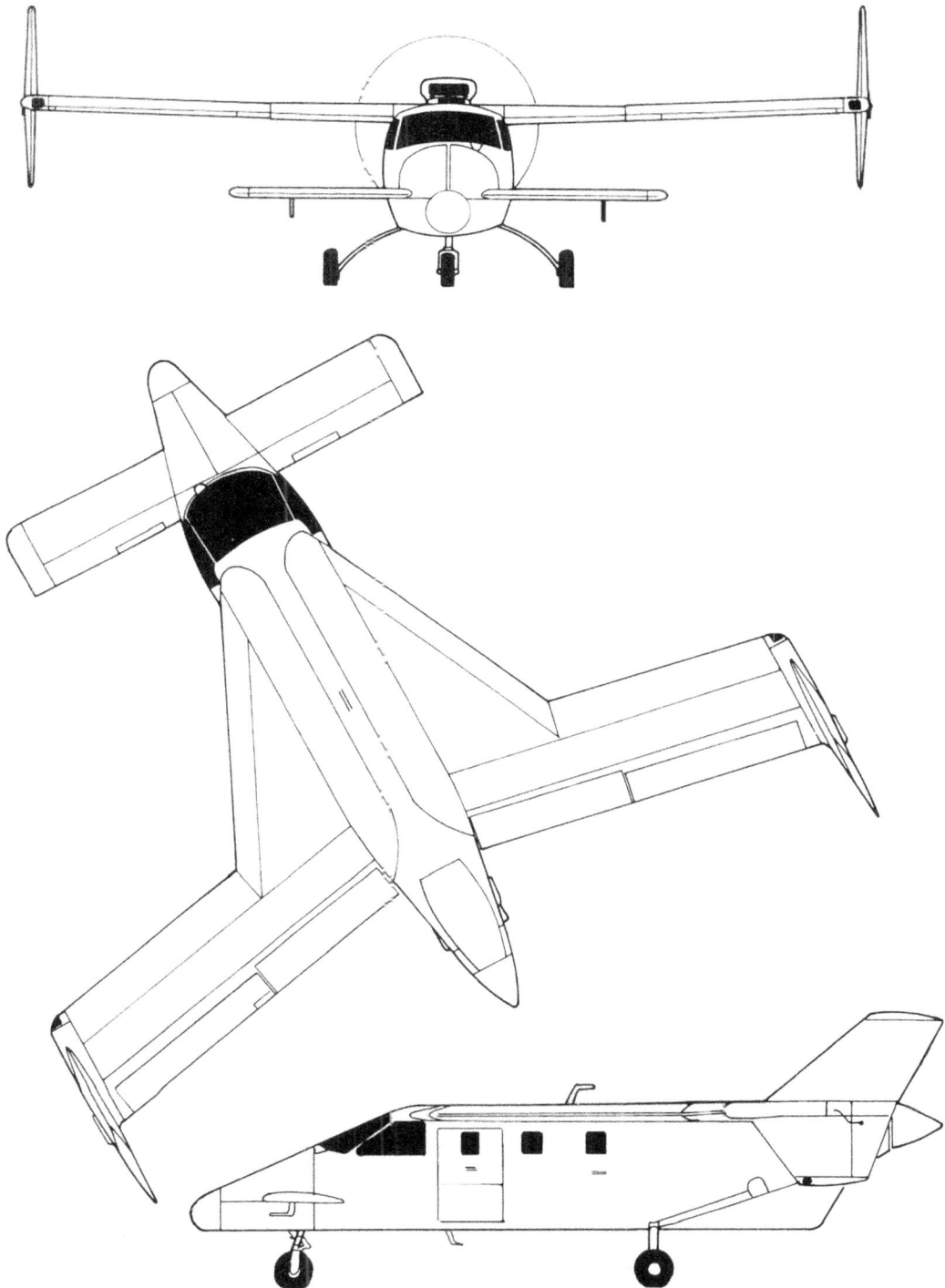

Fig. 4-12. Three view drawing of the OMAC-1.

winglets, providing directional stability and rudder control. The wing has 0° incidence, 1.5° dihedral, and no twist. Fuel is carried on the delta strake portion at the CG.

Ailerons outboard and flaps inboard cover the wing trailing edge. Kreuger flaps are employed on the portion of the wing leading edge outboard of the delta. These high lift devices are designed to lower the stalling speed to 70 mph, which the FAA currently requires of single engine airplanes.

The canard is 17 feet in span and carries the elevators; which, in addition to controlling pitch in normal flight, must also correct for the nose-down pitch resulting from flap deployment. Safety would dictate that aft wing flaps and canard elevator be linked for simultaneous trim action, when the flaps are lowered.

Initial flight tests were not satisfactory. Cable control to the ailerons permitted them to flex upward, reducing lift and disguising the fact that canard lift was inadequate. Use of rigid rods caused a heavy nose down action. The ailerons had behaved like lift increasing flaps, and both pilots had to apply stick pressure forcibly to stay level. A controlled crash with some damage resulted, since the right main landing gear leg did not come down.

As a result of this experience and further test flying other modifications were made:

- Canard Incidence is now 3°, instead of 1.5°
- Canard trim tab surface area has been increased 200%.
- Stainless steel torque tubes activate the trim system.
- Aileron differential was incorporated—upgoing aileron travel is now trice that of the going down unit.
- Winglets were increased in area substantially and the rudder cable control linkage was improved to eliminate cable stretch. The problem of slight yaw instability 5° either side of neutral persisted, however. Latest photos indicate addition of a small central fin on top of the engine air intake cowling, as a corrective measure.

The OMAC I is the first canard submitted for FAA certification. The basic design has been approved and that federal agency has not imposed any unusual requirements for flight or ground testing.

GATES-PIAGGIO GP-180

This "hybrid" aircraft is the result of extensive research, including wind tunnel testing in the Boeing Transonic Tunnel at Seattle. It is a joint effort between Gates Learjet of the United States and Piaggio of Italy. The aircraft was first announced at the NBAA Convention in Dallas on October 4, 1983, and the first flight was planned for April 1985.

The GP-180 is powered by two Pratt and Whitney PT6-A61 turboprop engines, flat rated at 700 shp for takeoff. They are mounted in shapely nacelles in the wings, driving pusher propellers.

The wing is unswept, of high aspect ratio, and carries ailerons and Fowler flaps. Its main spar, rear pressure bulkhead and main landing gear mounting are designed as a single component. Wing span is 45 feet-5 inches, and the area is 170 square feet. The forward wing has a slight anhedral, and an area of 23.59 square feet, including trailing edge flaps.

Directional stability and control is provided by a swept fin and rudder (and two ventral fins) topped by a swept T-tail with conventional elevators. The official Gates/Piaggio explanation of the principles behind his "hybrid" (author's terminology) configuration for takeoff, flight and landing is shown. This departure from the standard canard layout appears to make good sense.

One or two pilots and seven passengers are accommodated. A baggage compartment of 42 cubic feet is located behind the pressure bulkhead.

Tricycle landing gear retracts into the fuselage, which is pressurized to a 9. psi differential.

Performance figures are as follows: a maximum speed of 460 mph, a maximum cruise of 400 mph, a long range cruise of 368 mph, a rate of climb (two engines) of 3,650 fpm, and a rate of climb (one engine) of 1,250 fpm. Range with four passengers is 2,415 miles (with IFR reserves). Fuel capacity is 400 gallons. The ceiling is 41,000 feet. The aircraft has an empty weight of 6,400 pounds, a maximum weight of 9,800 pounds, and a useful load of 3,400 pounds.

Fig.4-13. The Gates-Piaggio GP-180 is a resurrection of the early Twentieth century hybrid concept, which features both a canard and a conventional tail. The joint effort between Gates Learjet of the United States and Piaggio of Italy, accommodates two pilots and seven passengers, and has forty-two cubic feet of baggage space. It is a mixture of composite and metal construction.

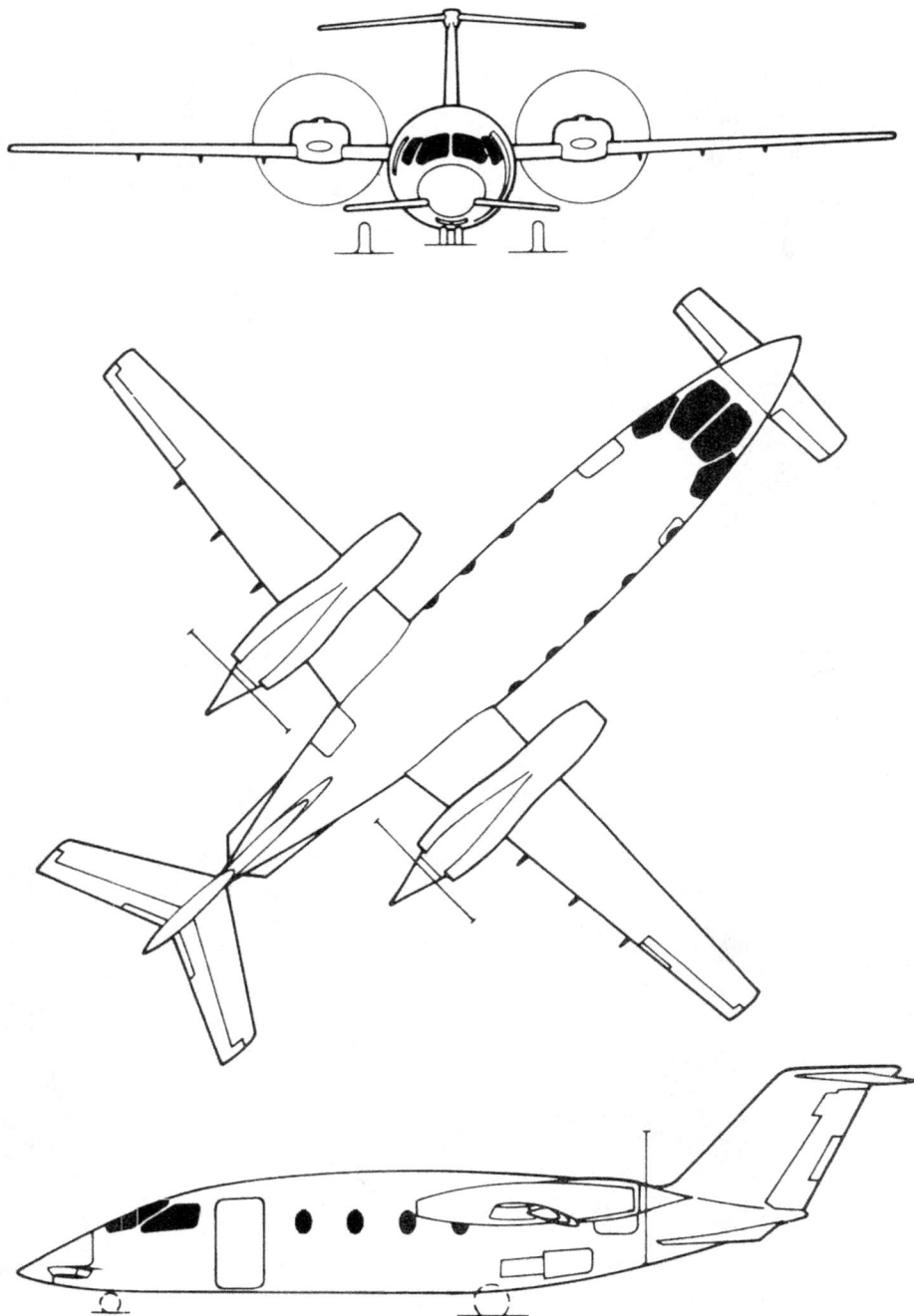

Fig. 4-14. Three view drawing of the Gates-Piaggio GP-180.

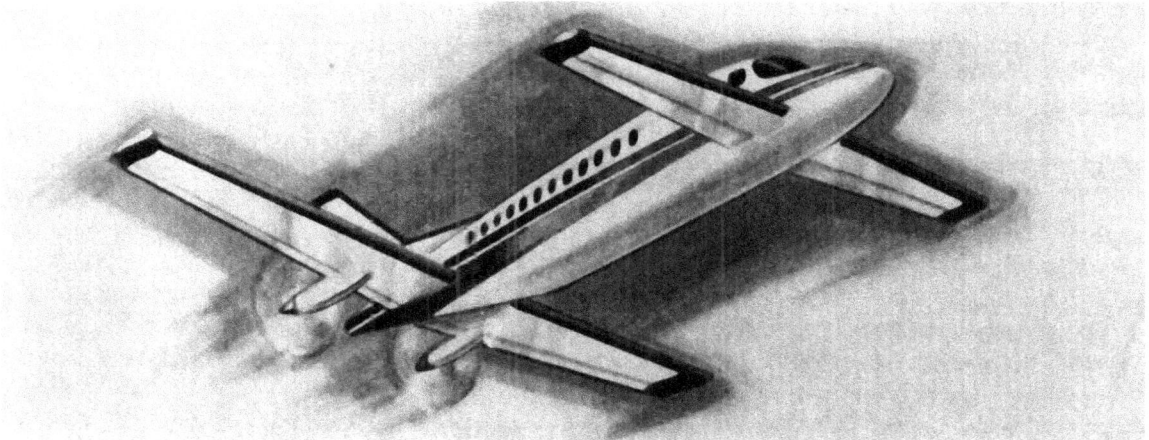

Fig. 4-15. Malcolm S. Harned of Cessna Aircraft Company, predicted this fifty passenger short-haul tandem wing airliner for the 1990's. Courtesy Cessna Aircraft Company.

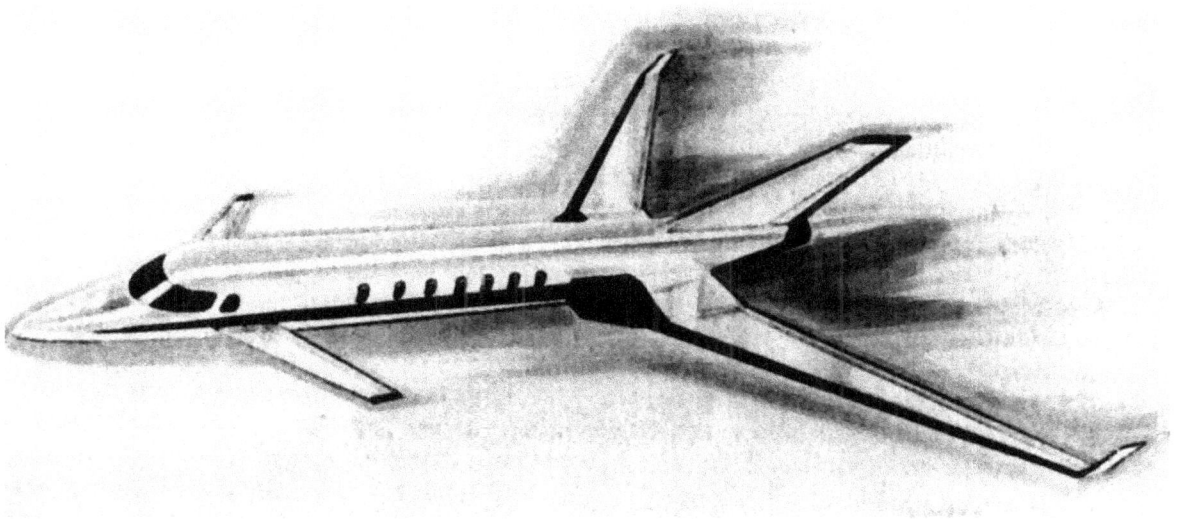

Fig. 4-16. Another Harned projection was this Mach 0.95, sixteen passenger business jet of canard configuration. Courtesy Cessna Aircraft Corporation.

HARNED'S PREDICTION

Malcolm S. Harned was senior vice president technology and a member of the board of Cessna Aircraft Co. Regrettably, he is now deceased.

In 1979, Mr. Harned wrote an article entitled "General Aviation Aircraft for the 1990's," which was published in the January 1980 issue of *Astronautics and Aeronautics,* and which has proven remarkably prophetic. His words reflected his own thinking; the ideas and projections in his article "do not represent the current official position of Cessna Aircraft Company."

In his words, Mr. Harned prophesied two types of aircraft, both canards:

The first was a short-haul, 50 passenger commuter. He visualized that the market for this type of aircraft would grow several fold, as it has already done with de-regulation of the airlines in the United States. It would have tandem wings to minimize trim drag, and for good control power for low takeoff and approach speeds. Aft turbo-prop pusher engines would permit a pleasantly quiet cabin which would be pressurized to cruise at 25,000 feet, where it could reach 300 mph, with an economical 100 seat miles per gallon of fuel.

The other canard was a MACH 0.95 business jet designed to fly 20% faster than today's jets, but with a higher fuel efficiency of four miles per gallon. Achieving this performance called for engines buried in the fuselage, an area ruled fuselage, highly swept wings, and super critical airfoils. Canards would minimize trim drag, and winglets would do double duty, increasing effective aspect ratio, as well as providing for directional stability. The aircraft would cruise at 600 mph at altitudes up to 60,000 feet, carry 16 people, and its headroom would permit occupants to stand erect in the aisles.

He foresaw the weight savings from composites, high power-to-weight ratios, and an improved specific fuel consumption for turboprop engines.

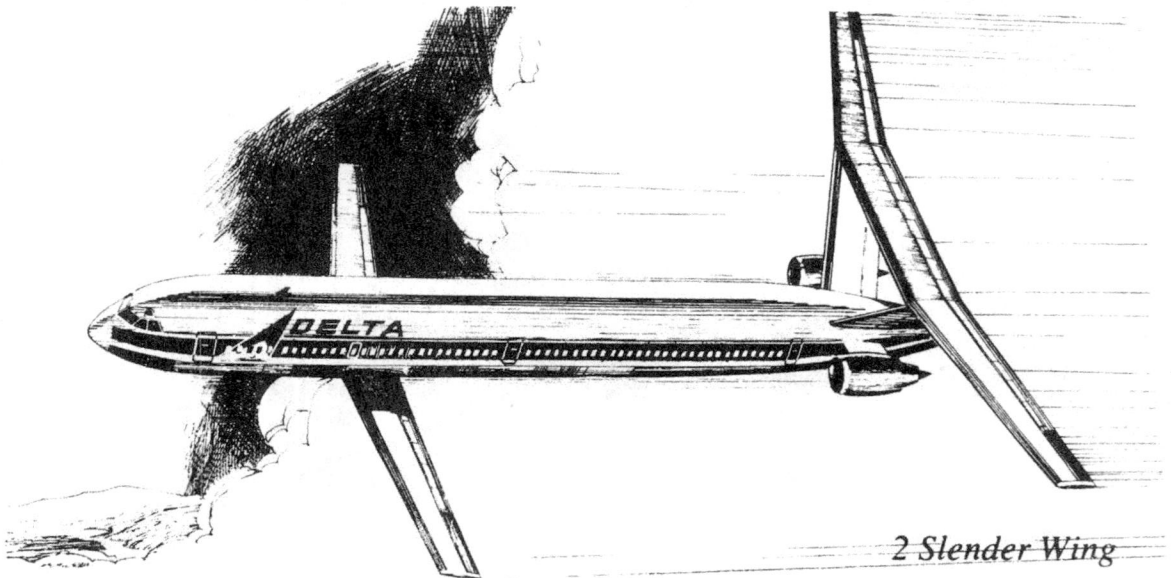

Fig. 4-17. A proposal for a Delta Airlines tandem wing airliner.

14

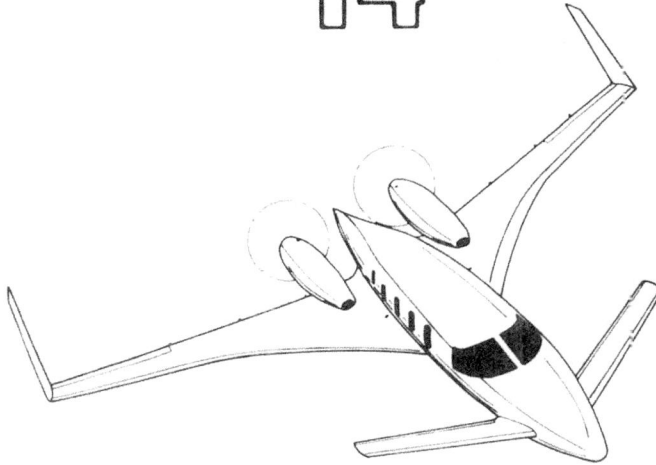

Amateur-Built Canards

The development and acceptance of the modern canard can be attributed directly to the homebuilt aircraft movement. Here's a look at what the future may hold for amateur aircraft builders.

JONAS "HUMMINGBIRD"

This small two place aircraft has not, to the author's knowledge, been flown, and should be regarded as a "project-in-process." The Hummingbird is a unique combination of sophisticated aerodynamic and structural features and could only be flown by a pilot experienced with such things as flaps, speed brakes, etc.

The wing spans 20 feet, while the fuselage is 10 feet long. Total wing area is 41 square feet, of which the aft wing is 32 square feet and the canard nine square feet. Based on the wing area alone and neglecting the lift contribution of the airfoil sectioned fuselage, the wing loading is 20.73 pounds per square foot and the power loading is 10.89 pounds per horsepower.

Empty weight is 435 pounds, while gross weight is 850 pounds. The power plant a Revmaster 2,100 cc VW conversion, developing 78 horsepower at 3,600 rpm. It turns a Maloof two speed aluminum pusher propeller, 56 inches in diameter.

Speeds are estimated at 240 mph top, 220 mph cruise, and a stall of 64 mph. Landing speed would be in the 70 mph range presumably with flaps and spoilers deployed.

Because of its small size, the wing chords are narrow. The canard chord measures in at 15 inches while the wing root is 30 inches and the tip 18. Consequently, these surfaces will be operating at low Reynolds numbers. At a landing speed of 70 mph, the canard Re would be 819,000 at sea level.

The fuselage is almost rectangular in top view and 47 inches wide at the maximum. Occupants sit side by side in a "two head clearance bubble" canopy, giving a feminine cross section in front view. The canopy opens at the front, and since it contributes to lift, must be firmly latched in flight.

The fuselage ends rather abruptly. Cooling vents on the top surface and rear cooling air

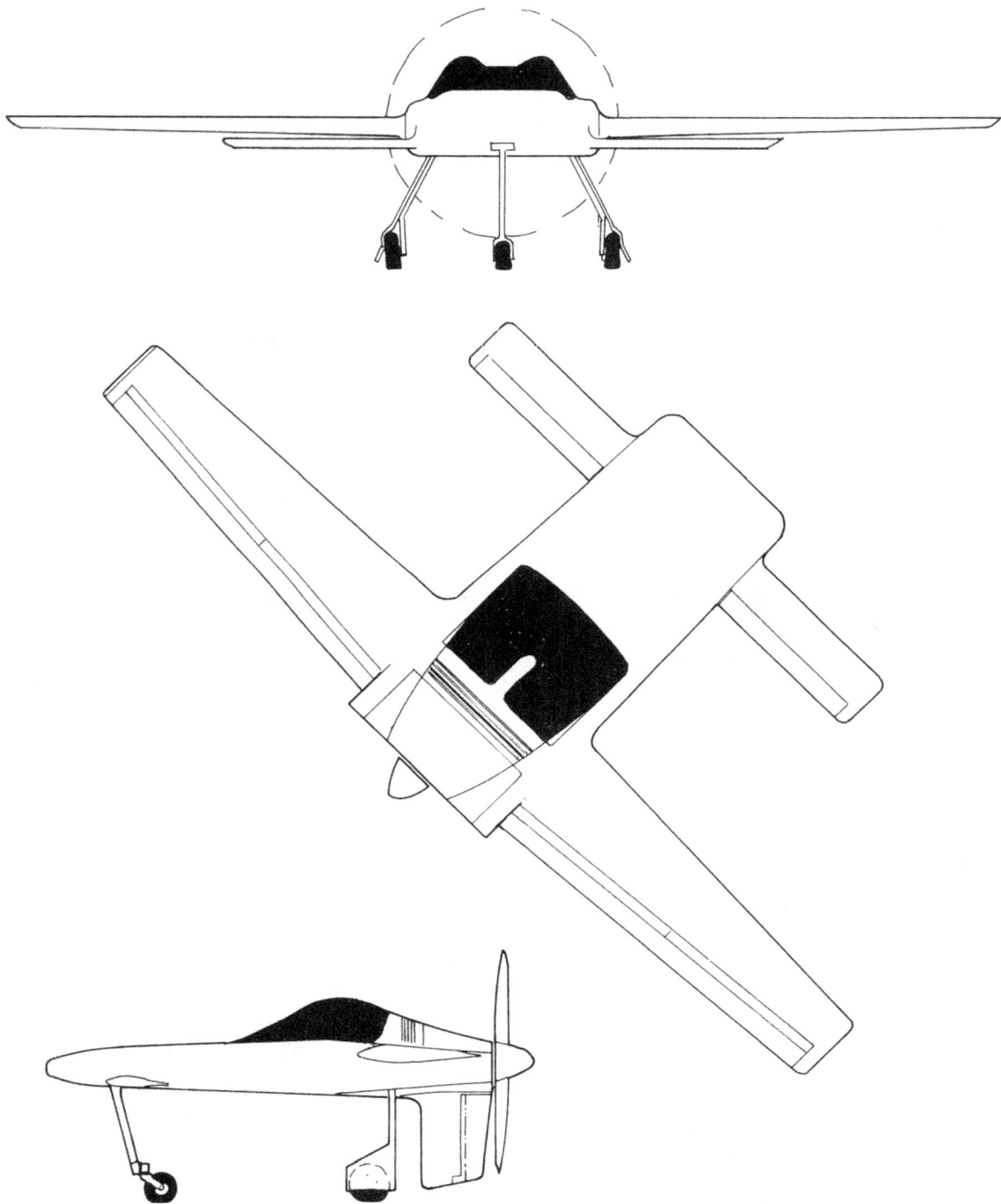

Fig. 4-19. Three view drawing of the proposed Jonas Hummingbird, high performance, amateur-built canard. It is designed to carry two people at an estimated 220 mph cruise.

exits, permit the propeller to function as a suction pump. Sucking in boundary layer air via the vents is claimed to reduce drag.

The foreplane uses the NASA 64-412 airfoil set at 0° angle of incidence. The elevators are 27% of the chord, being the slotted flap type. Deflection is down 45° and up 15°. For landing, these must be coordinated with the aft wing flaps by hand. The aspect ratio of the canard is seven.

The aft plane's airfoil section is NASA 642-215. The wing tapers about its 25% chord, and the taper ratio is root five tip three. The upper surface is flat across and the taper gives 2° dihedral on the under surface. There is no twist in this aspect ratio 12 wing.

High lift devices on the aft wing are full span, reflexible 20% chord plain flaps. Roll control is achieved by spoilers, outboard. Inboard, they serve as speed or dive brakes, and are separate from the outboard spoilers. Both are located at 70% of chord from the leading edge. The spoilers give positive yaw and are particularly effective at low speeds, with flaps deployed, similar to the slot-lip ailerons.

The speed brakes act in concert with the flaps, to prevent pitch changes and provide considerable drag. Thus, while the flaps can normally only be deflected 10° to 15° in combination with the elevator, an additional deflection of 10° to 15° is obtainable with the speed brakes. This implies

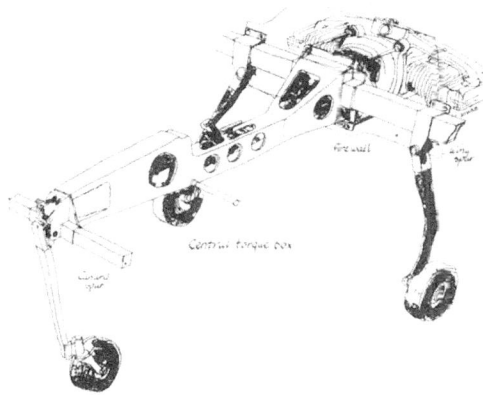

Fig. 4-20. Drawing of the central torque box structure of the Hummingbird.

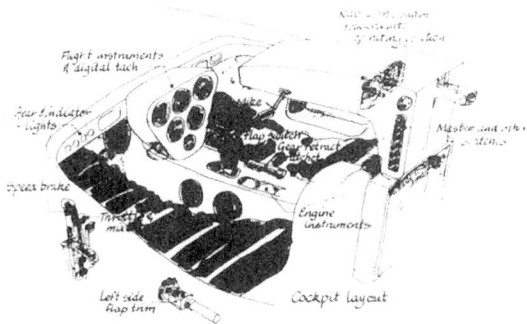

Fig. 4-22. Artist's rendering of the Hummingbird's control systems.

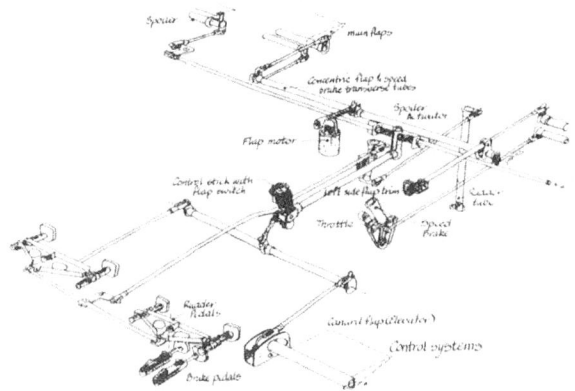

Fig. 4-21. The Hummingbird's cockpit layout features a center console, and supine seating.

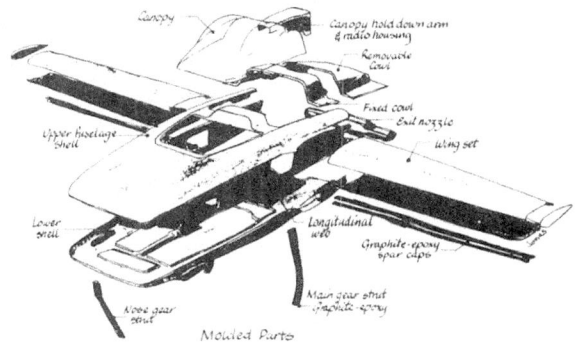

Fig. 4-23. An exploded drawing of the Hummingbird's molded parts suggests that goes together like a plastic model airplane kit.

that the flaps must be deployed before the dive brakes can be used. Deployment of the flaps on both canard and wing, plus the spoilers, should permit steep, slow, descents, and rapid deceleration on the runway. The flaps may also be reflexed for trim control.

The ventral fin and rudder location is quite unusual on a small aircraft. A small bumper wheel limits flare to about 7°, and prevents the propeller from ground contact, when the plane is empty. This position allows the rudder to act like an aileron, giving proverse roll. It helps the aircraft roll in the intended yaw direction, unlike the more normal dorsal rudder, which wants to roll the aircraft opposite to the yaw. Another reason for the bottom vertical is that in high angles of attack, a top vertical could be blanketed by the wide fuselage. The rudder has a high aspect ratio and uses the NACA 0012 section at the top, and 65-015 at the bottom.

Fuel tanks run along both sides of the fuselage with each side containing ten gallons. A small gravity feed tank, just behind the cockpit, holds ¼ gallon. Cruise fuel consumption is 3.5 gallons per hour. The centers of the tanks coincide with the CG, and they are baffled to prevent a rearward destabilizing fuel surge in an extended climb.

The gear is long and the main wheels overlap,

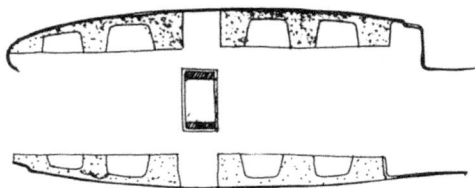

Fig. 4-24. Exploded cross section of the Hummingbird's wing reveals its composite construction.

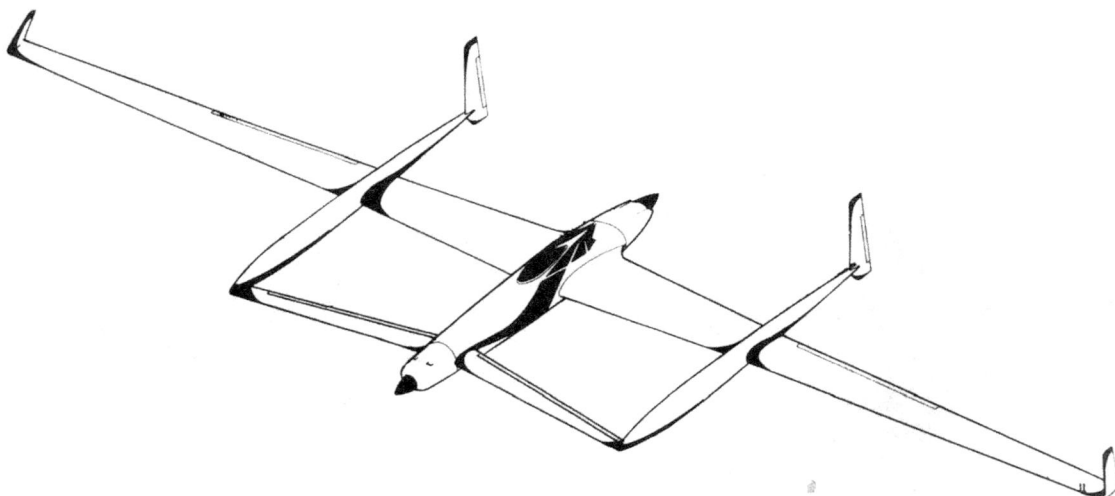

Fig. 4-25. Artist's rendition of the Rutan Voyager. The 110 foot span, twin-engine two-seater was designed to fly around the world non-stop, non-refueled. The trip is planned to take ten days for completion. It is designed to carry twelve times its weight in fuel.

requiring a special geometry hinge axis, with one wheel retracting well ahead of the other. The main wheels have dual disc brakes, facilitating on-the-ground steering. The nosewheel is a castering retractable, and is fitted with anti-shimmy damping.

All three wheels are mechanically retracted from a single ratchet, mounted on the central box, via a torque tube gear box and jack screw arrangement. Note that the aft wing flaps are electrically operated from a switch on the central control stick. This permits prompt elevator deflection to offset the severe nose down pitch change from the deflected flaps. Also note that the left flap has trim for single occupant trimming. Only the spoiler on the inside of the turn lifts for roll control. The underslung rudder is advantageous from the control linkage point of view, but a wheels-up landing would be "bad news."

The molded components are made in female molds following European glider practice. While this demands expensive molds, the resulting components have very smooth exterior surfaces requiring very little finishing.

Further information on this intriguing little canard can be obtained from Jonas Aviation, 1355 Van Dyke, San Francisco, CA 94124.

THE RUTAN VOYAGER

The Voyager is a state-of-the-art, composite construction canard, with autoclaved carbon spars and advanced sandwich skins. It is designed to carry a crew of two, with provision for about ten days.

The configuration is a virtual breakthrough in that it solves the tremendous aeroelastic problems associated with carrying fuel weighing twelve times the 110 foot span aircraft's empty weight. The airfoils were designed by John Roncz,

Fig. 4-26. The SCALED prototype ARV designed by Burt Rutan is a portend of things to come in this exciting segment of aviation.

optimized for the required performance, in consideration of smashed insect carcasses cluttering the wings. Needless to say, extreme care was taken in optimizing strength-to-weight ratios for all systems, to allow the lightest, strongest structure possible:

Two years of intensive effort by crewmembers Dick Rutan and Jeana Yeager, and others of the Rutan staff, resulted in the completion of the amazing aircraft in May of 1984, Flight testing was expected to take place in the summer of 1984, but without the modified engines and optimized propellers required for the global assault. The two engines are actually required in order to match power requirements to the drastic weight changes which will occur during the flight. Spec-

ialized, expensive global navigation equipment will not be installed until the aircraft is ready for the record attempt.

CATTO ACRO-X

Looking like something out of the movie "Star Wars," the Acro-X is actually a composite construction canard with cable-braced wings. It was put on static display at the 1983 ARV competition in Oshkosh, but was not flown, because its restrictions had not been flown off. No word on its further development was available at press time, but the concept appears to have some merit.

The primary specifications include a wingspan of 16 feet-4 inches, and empty weight of 300 pounds, and a cruising speed of 105 mph.

Fig. 4-27. The Catto Acro-X is truly a blend of old and new concepts—a wire-braced biplane with a canard, and constructed of composites.

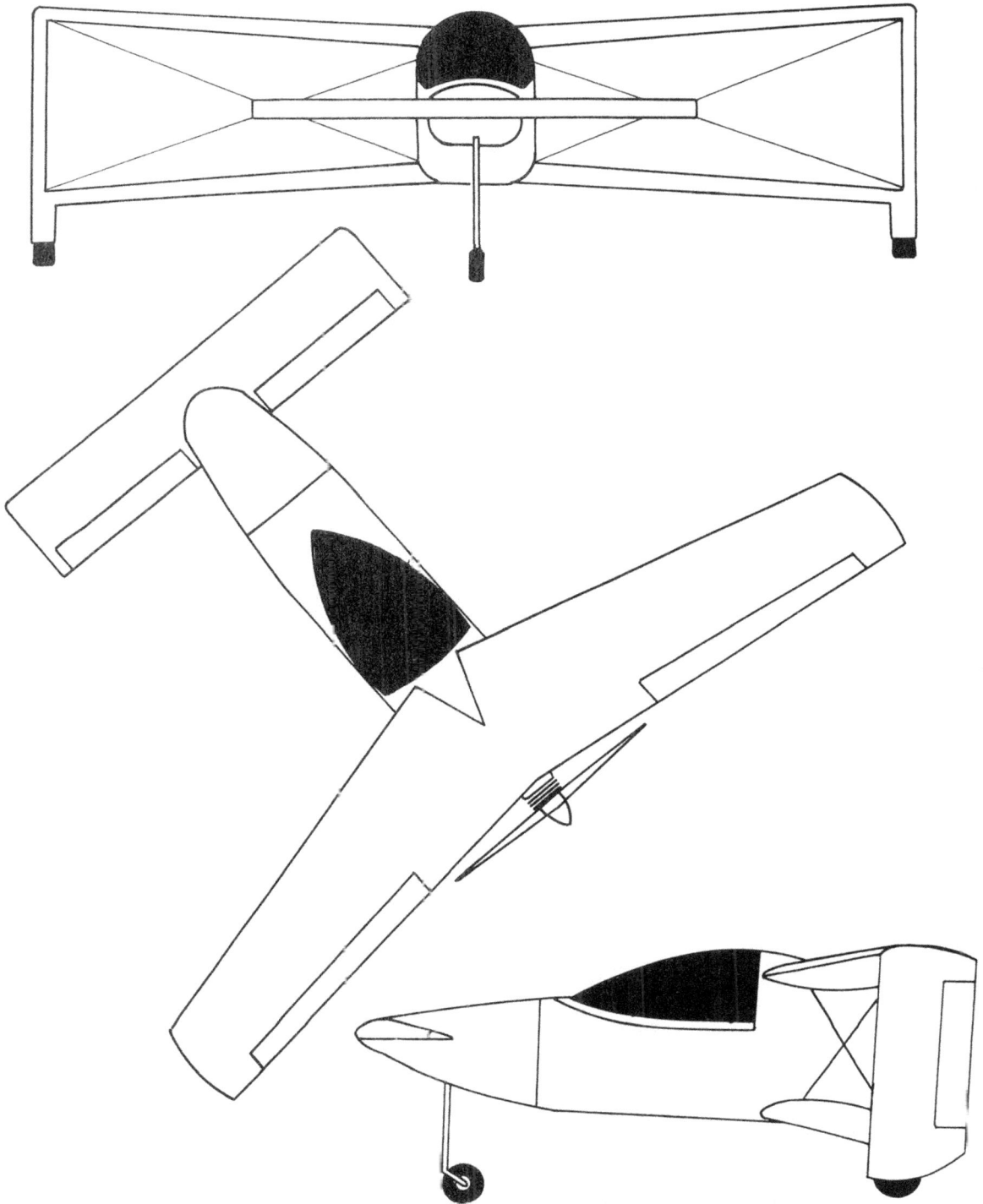

Fig. 4-28. Three view drawing of the Catto Acro-X.

15

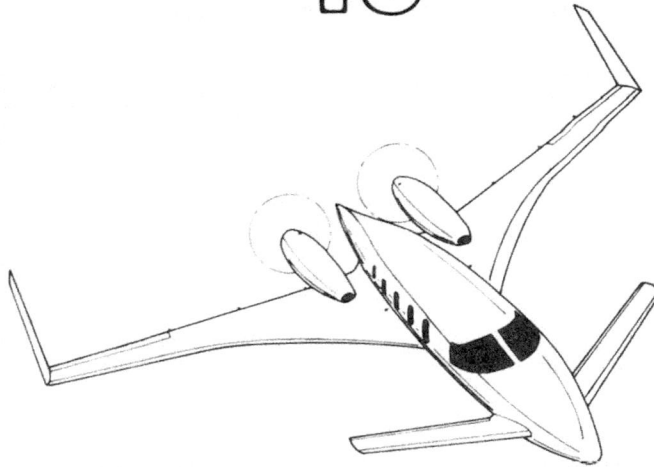

Fighter Aircraft Canards

Today's costs of designing, testing and building military aircraft are astronomical. In addition, the life-cycle cost of operating and maintaining these aircraft throughout their service life, adds enormously to the costs of National Defense. Because of this, smaller nations, such as Great Britain, France, Italy and Israel are combining their forces to develop joint projects, not only for the aircraft, but also the power plants. Since, generally, the heavier aircraft is the more expensive, keeping size and weight down, while fulfilling realistic mission requirements, is the challenge.

Advanced titanium and composite materials are beginning to dominate future weapons systems, as well as general and transport aviation. This is the most significant breakthrough since aluminum stressed skin construction replaced fabric, wood, and steel tube structures. In addition to the weight/cost reductions that these new materials permit, automation to reduce the cost of labor intensive operations is growing rapidly, in this age of computers and industrial robots.

The computer is entering the cockpit. Digital flight control systems are replacing today's aerodynamic stability criteria, resulting in a wide range of new agilities, and combat maneuvers previously not attainable due to stability limitations. Data processing using micro processors, allied with fly-by-wire (no direct mechanical linkage of controls to flying surfaces) guidance systems, and quick responding, powerful actuators for moving heavily loaded control surfaces, are becoming prevalent.

Super-critical airfoils are being developed to permit higher wing loadings, and greater lift and maneuverability with less drag. All are combining to reduce aircraft weight, size and cost. Control-configured-vehicle (CCV) technology will also be incorporated in new aircraft designs. In addition to the increased maneuverability, smaller, lighter control surfaces, with less drag, are possible.

The close-coupled canard, with either Delta wings or swept forward wings (e.g. Gruman X-29),

Fig. 4-29. The Grumman X-29a is a prime example of the coming age of advanced fighterjets. It features a fully flying canard capable of rotating through 60-degrees, and a forward swept wing. It incorporates a relaxed longitudinal stability, and a triply redundant fly-by-wire flight control system. The wing is constructed of composites, while the remainder of the aircraft is conventional aluminum and steel.

is being developed. All this increased flight activity will impose greater strains on pilots' physiques, calling for seats that tilt back to reduce G loads on the crew, and better G suits, etc.

The swept forward wing has seen previous use, principally in German designs. Both aft and forward sweep impose high twisting loads on the wings. The twist in aft-swept wings tends to reduce wing tip incidence; but in forward sweep, it induces an increase in incidence that exagerates the twist, leading to failure. A swept forward wing must have much greater torsional rigidity. Through its composite research, Grumman has overcome this major problem, permitting them to capitalize on the aerodynamic advantages of forward sweep.

Aft swept wings tend to stall at their tips at high angles of attack, which also reduces aileron effect-iveness. Forward sweep does just the opposite. The wing roots stall first, leaving the wing tips and ailerons effective for full roll control and spin resistance. As a bonus, supersonic performance is also improved.

Swept forward wings must be placed further aft on the fuselage. Therefore, it is advantageous to place a canard ahead of the wing. Close-coupling of canard and wing, also permits directing the airflow over the aft wing root, resisting root stall. The tip vortex from the canard energizes the airflow over the wing, delaying the stall.

On the Gruman X-29, the canard is all moving. Instead of the CG being ahead of the center of lift of the aft wing, balanced by canard lift, the CG is aft—the aircraft is aerodynamically unstable and relies on computer control to stay in flight. Reduced control surfaces are possible, leading to less weight. Less weight and drag require less power

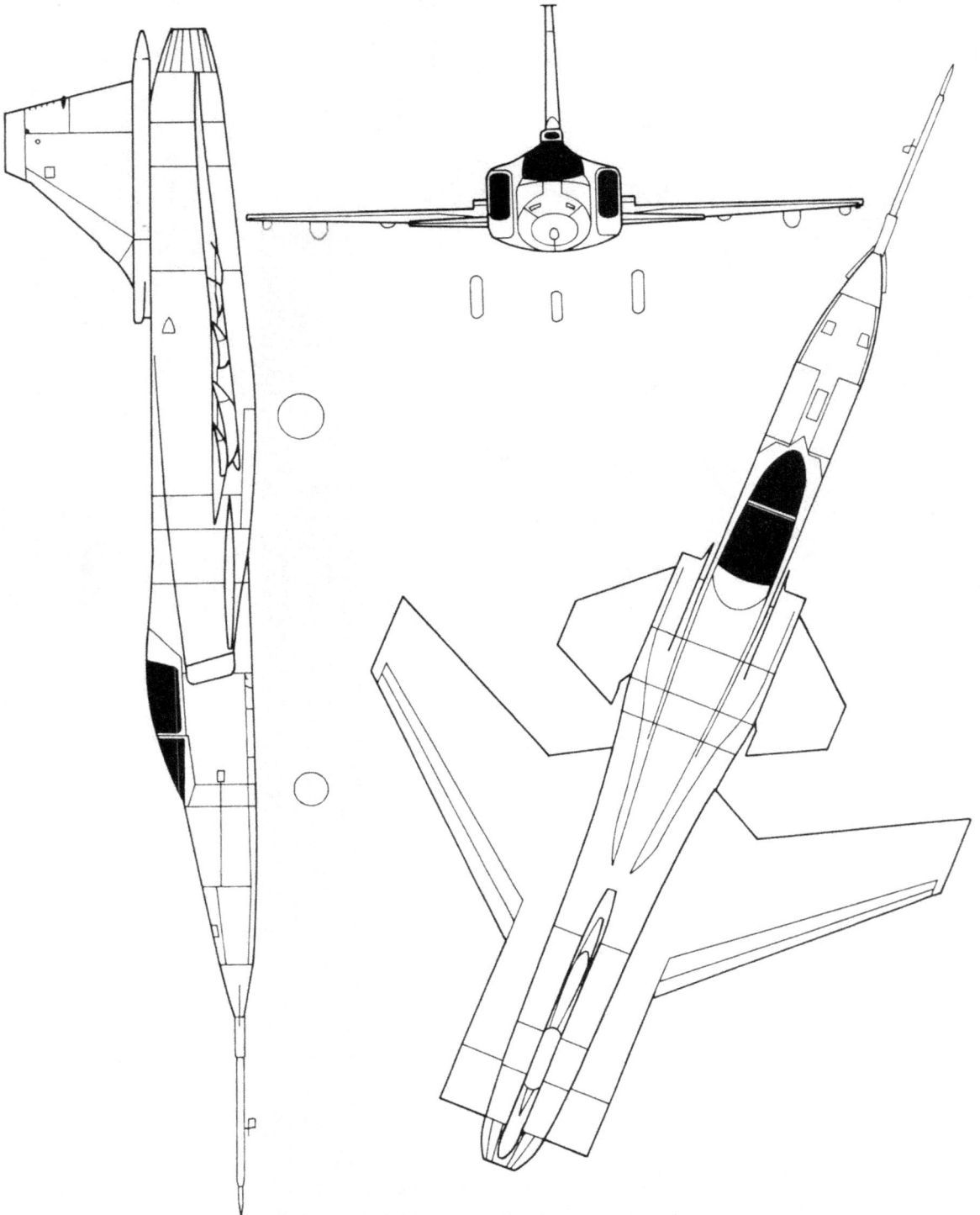

Fig. 4-30. Three view drawing of the Grumman X-29a fighter.

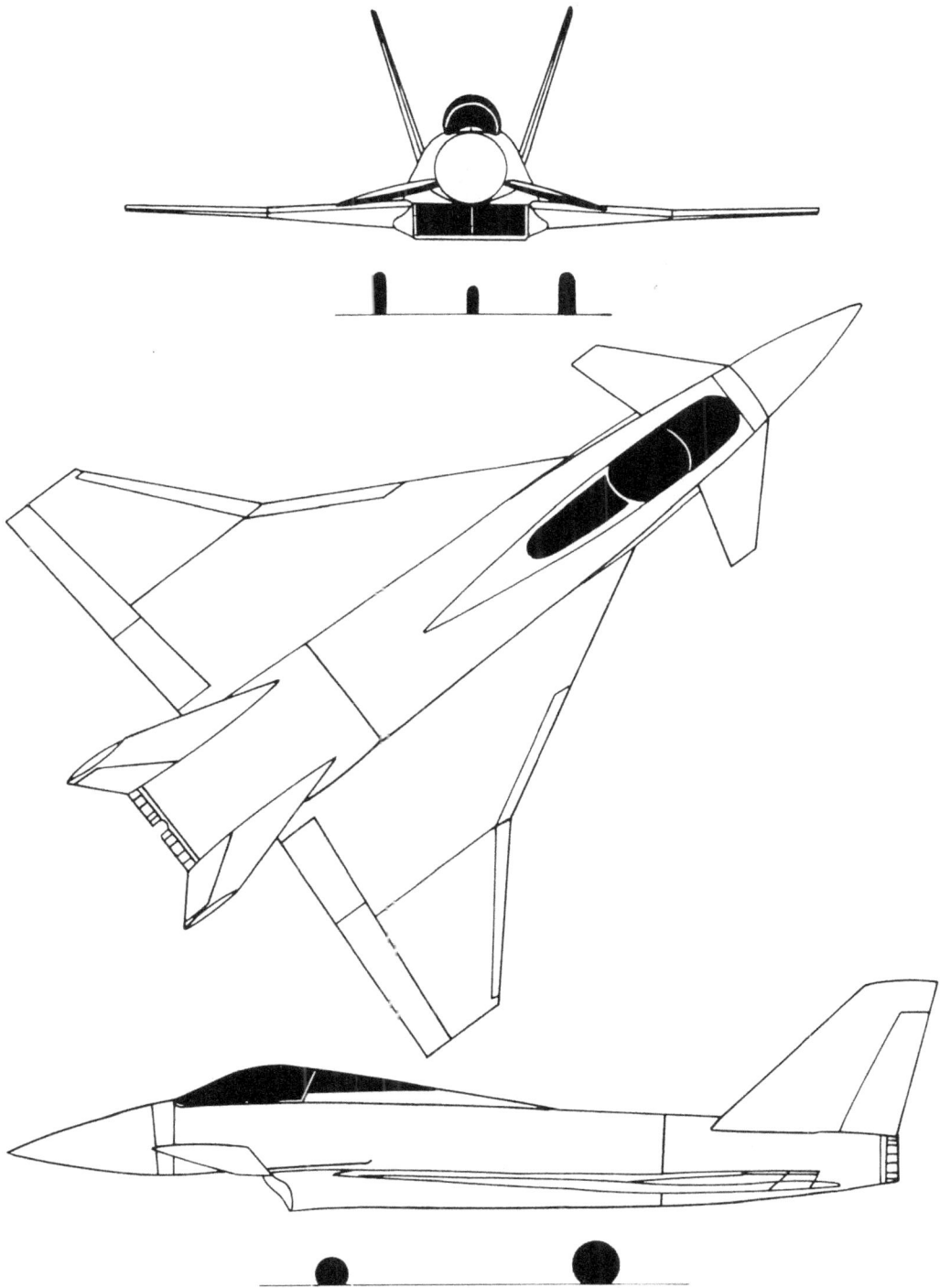

Fig. 4-31. The AGA (Agile Combat Aircraft) by British Aerospace, MBB, and Aeritalia.

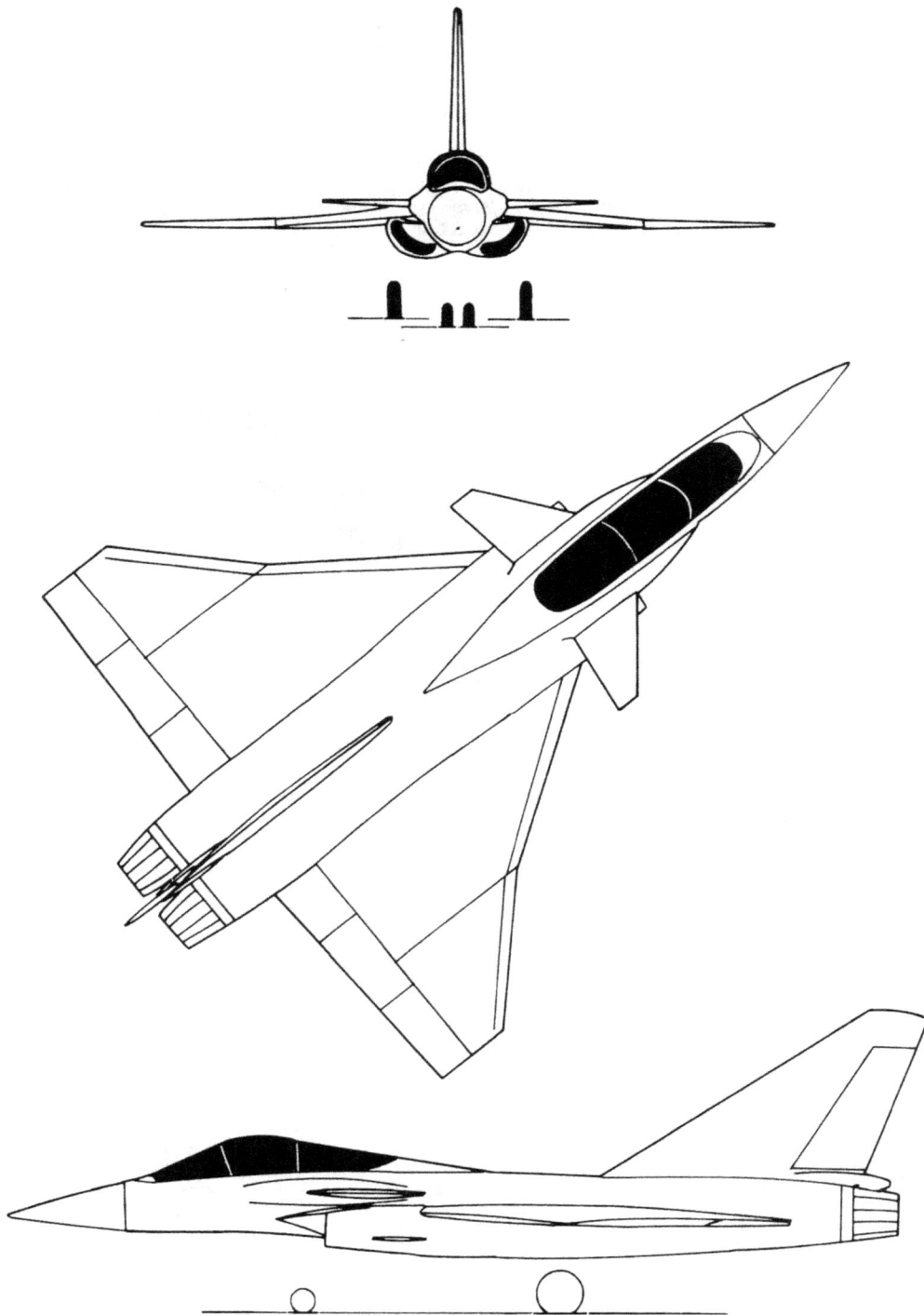

Fig. 4-32. The Aeritalia and Dassault-Breguet fighter.

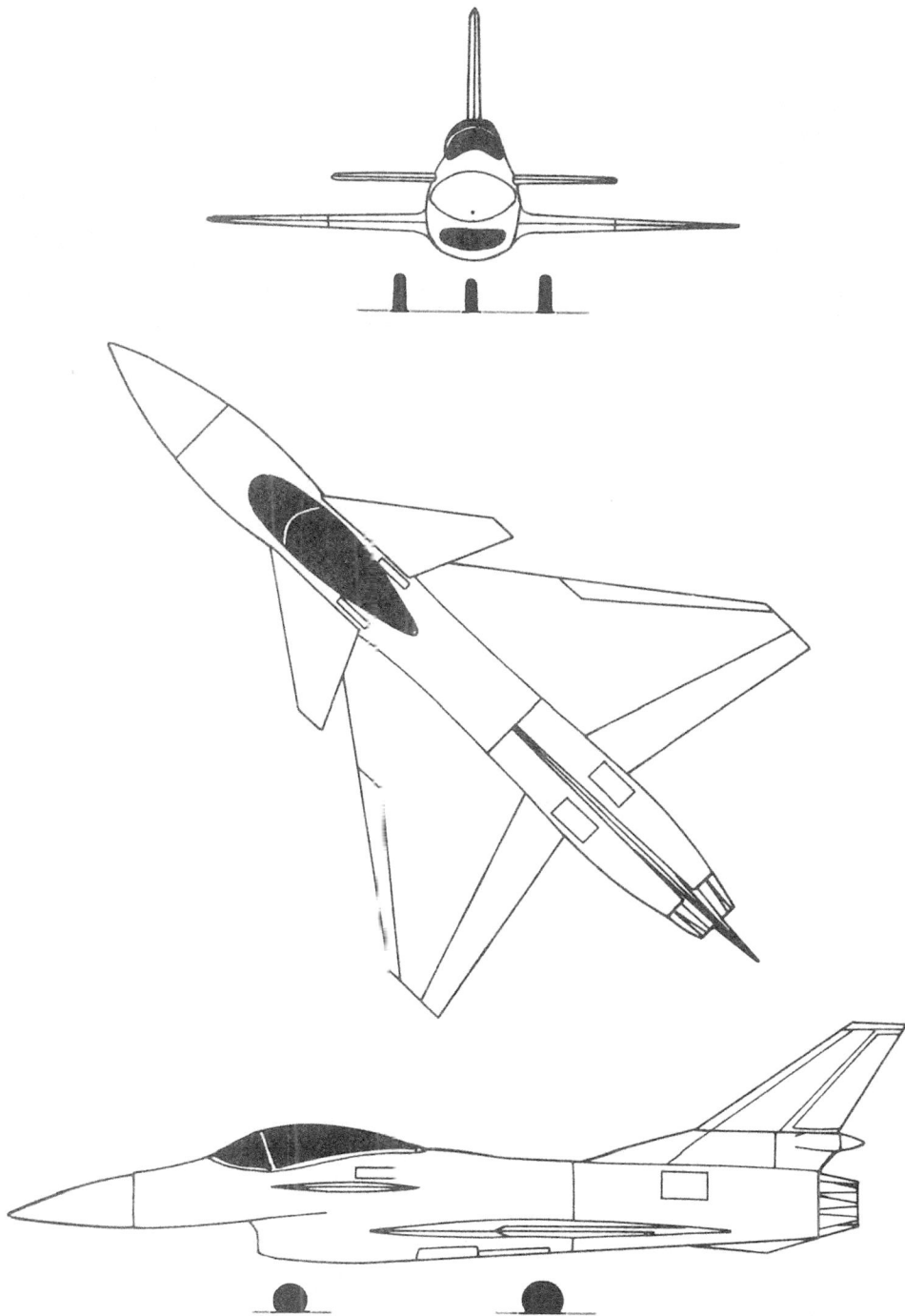

Fig. 4-33. The IAI Lavi, Dual-Role Fighter, planned for service in 1989.

and less fuel, reducing both first and service life costs.

Besides the U.S. Developments, other nations are developing CCV's but employing delta aft wings. British Aerospace has their A.C.A., and Dassault-Breguet of France their A.C.X.; Israel has its Lavi, developed from Dassault technology. Sweden's SAAB is developing the J.A.S. 39 Gripen, multi-roll combat aircraft, as a follow-on to their Viggen. They are all planned for 1990's operation.

The artificial stability that computerized control surfaces permit, can be utilized to perform previously impossible maneuvers. Evasive, wings level, side slips; flat, unbanked turns, and yaw-pointing, can be achieved. In addition, while still level, the aircraft can lift upward or move downward or sideways, by unorthodox use of canards, elevons, and rudder.

Redundant control systems are obviously needed, since failure of a single system could only lead to disaster. Human control unaided by a computer, is not possible.

Interestingly, the Boeing Aircraft Co. is currently testing active controls (A.C.T.). This system incorporates pitch axis stability augmentations, fly-by-wire pitch axis control and wing load alleviation. It also includes maneuver load control and gust alleviation control. The wing is moved forward to a position of negative stability which is compensated for by active controls. The horizontal tail (of a conventional form) is reduced 45% in area for both weight and drag reduction, providing improved performance at lower fuel consumption. Triple redundant systems will be incorporated for safety.

Appendix

A-1

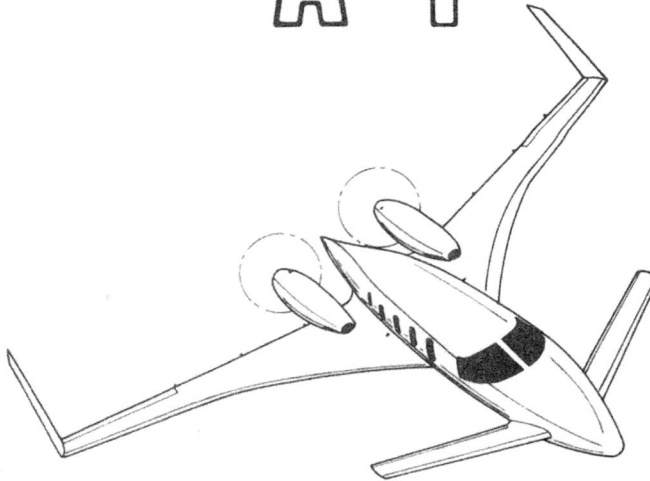

References And Their Sources

A listing of forty references follows.

In Canada, Municipal Libraries have an inter library loan system that makes it possible to obtain short term loan of texts or copies of reports, technical notes, etc. at very low cost.

In essence, membership in a Canadian Municipal Library provides entry into other Canadian Municipal Libraries, into University Libraries and into the Canadian Institute for Scientific and Technical Information Library. The writer has been amazed to successfully obtain some old and obscure British and German (translated into English) reports through this system.

In the U.S.A., the U.S. Department of Commerce, National Technical Information Service, 5285 Port Royal Road, Springfield, Virginia 22161 U.S.A. can supply data when ordered, specifying in detail what is required, including authors. Also try your local Municipal Library; University Library, or Library of Congress, Washington, D.C. Another U.S. source is Aero Space Research Applications Center, 611 N. Capitol Avenue, Indianapolis, Indiana, 46204. Try Zenith Aviation Publications for texts.

In this listing of references, the title is given first solely for the readers' convenience. This is not intended to diminish, in any way, the enormous contribution to aviation knowledge made by the dedicated scientists who authored these references.

At this point, the author wishes to acknowledge the great assistance that "Sources Public Library," Roxboro, Que. has provided in obtaining texts and copies of technical reports; particularly in the gracious persons of Mrs. Norah LaRue and laterally Mrs. Claudette Fournier.

1. AIRCRAFT PERFORMANCE STABILITY
   AND CONTROL:
       Perkin & Hage
       Published by John Wiley & Sons Inc.

2. THEORY OF WING SECTIONS:
       Abbott & Von Dornhoff
       Published by Dover Publications Inc.

3. AERODYNAMICS FOR NAVAL AVIATORS
   (NAVWEPS 00-802-80
   HURT (U.S.C.)

Issued by the Office of the Chief of Naval Operations
Aviation Training Division

4. NACA Report No. 586 —
   AIRFOIL SECTION CHARACTERIS-
   TICS AS AFFECTED BY VARIATIONS
   OF THE REYNOLDS NUMBER
   Jacob & Sherman (1939)

5. NACA Technical Note 1945 —
   AERODYNAMIC CHARACTERISTICS
   OF 15 N.A.C.A. AIRFOIL SECTIONS AT
   SEVEN REYNOLDS NUMBERS FROM
   0.7 x 10.6 to 9.0 x 10.6
   Loftin & Smith (1949)

6. NACA Report No. 572 —
   DETERMINATION OF THE CHARAC-
   TERISTICS OF TAPERED WINGS
   Anderson (1940)

7. NACA Report No. 460 —
   THE CHARACTERISTICS OF 73 RE-
   LATED AIRFOIL SECTIONS FROM
   TESTS IN THE VARIABLE DENSITY
   WIND TUNNEL
   Jacobs - Ward - Pinkerton (1939)

8. NACA Report No. 407 —
   THE CHARACTERISTICS OF A CLARK
   Y WING MODEL EQUIPPED WITH
   SEVERAL FORMS OF LOW-DRAG
   FIXED SLOTS
   Weick - Wenzinger (1932)

9. NACA Report No. 628 —
   AERODYNAMIC CHARACTERISTICS
   OF A LARGE NUMBER OF AIRFOILS
   TESTED IN THE VARIABLE DENSITY
   WIND TUNNEL
   Pinkerton - Greenberg (1938)

10. NASA Technical Paper No. 1589 —
    EXPLORATORY STUDY OF THE E-
    FFECTS OF WING LEADING EDGE
    MODIFICATIONS ON THE STALL/
    SPIN BEHAVIOUR OF A LIGHT GEN-
    ERAL AVIATION AIRPLANE
    Langley Research Center (1979)

11. NASA Technical Memorandum TMX 72697 —
    LOW SPEED AERODYNAMIC CHAR-
    ACTERISTICS OF A 13 PERCENT
    THICK AIRFOIL SECTION DESIGNED
    FOR GENERAL AVIATION APPLICA-
    TIONS
    McGhee et al (1977)

12. NASA Technical Note TND 7428 —
    LOW SPEED AERODYNAMIC CHAR-
    ACTERISTICS OF A 17 PERCENT
    THICK AIRFOIL SECTION DESIGNED
    FOR GENERAL AVIATION APPLICA-
    TIONS
    McGhee - Beasley (1973)

13. A.R.C. CP 1187 - 1971 —
    EXPERIMENTAL INVESTIGATION OF
    A HIGH LIFT LOW DRAG AIRFOIL
    (GU25-5(11)8)
    Kelling (1971)

14. NASA Technical Paper No. 1865 —
    DESIGN AND EXPERIMENTAL RE-
    SULTS FOR A FLAPPED NATURAL
    LAMINAR FLOW AIRFOIL FOR GEN-
    ERAL AVIATION APPLICATIONS
    Somers (1981)

15. FLUID DYNAMIC DRAG
    Hoerner (1965)

16. FLUID DYNAMIC LIFT
    Hoerner & Borst (1975)

17. DEVELOPMENT OF A SMALL, HIGH ASPECT
    RATIO CANARD AIRCRAFT
    Rutan
    Society of Experimental Test Pilots Symposium
    September 1976.

18. PROPORTIONING A CANARD AIRPLANE FOR
    LONGITUDINAL STABILITY AND SAFETY A-
    GAINST STALL
    Journal of Aeronautical Sciences PP 523-528
    December 1942
    Foa

19. ROYAL AIRCRAFT ESTABLISHMENT—
    FARNBOROUGH
    Technical Notes 1499 August 1944
    1687 September 1945
    Flight tests of the Miles Libellula Tandem Bi-plane
    Alston - Brotherhood - Ewans

20. ROYAL AIRCRAFT ESTABLISHMENT—
    FARNBOROUGH
    Report No. BA-1542 - July 1939
    Notes on Tail First Aeroplane - Gates

21. MINISTRY OF AIRCRAFT PRODUCTION RTP
    TRANSLATION No. 2107
    DEVELOPMENT AND PROBLEMS OF THE
    "CANARD" AIRCRAFT
    Helledoren 1941

22. DESIGN FOR FLYING
    David B. Thurston
    Publisher McGraw Hill, New York, N.Y.

23. NACA REPORT No. 648 —
    DESIGN CHARTS FOR PREDICTING
    DOWNWASH ANGLES AND WING
    CHARACTERISTICS BEHIND PLAIN
    AND FLAPPED WINGS
    Silverstein & Katzoff (1939)

24. NACA REPORT No. 664 —

WIND TUNNEL INVESTIGATION OF AN NACA 23012 AIRFOIL WITH VARIOUS ARRANGEMENTS OF SLOTTED FLAPS
Silverstein & Harris (1939)

25. NACA TECHNICAL NOTE No. 808 —
WIND TUNNEL INVESTIGATION OF AN NACA 23012 AIRFOIL WITH SEVERAL ARRANGEMENTS OF SLOTTED FLAPS WITH EXTENDED LIPS
Lowry 1941

26. AERONAUTICAL RESEARCH COUNCIL (Great Britain)
TECHNICAL REPORT No. 2622 —
AERODYNAMIC CHARACTERISTICS OF FLAPS
Young (1947)

27. NACA REPORT No. 738 —
GROUND EFFECT ON DOWNWASH ANGLES AND WAKE LOCATION
Katzoff and Sweberg (1943)

28. NACA REPORT No. 605 —
RESUME AND ANALYSIS OF NACA LATERAL CONTROL RESEARCH
Weick - Jones (1937)

29. NACA REPORT No. 868 —
SUMMARY OF LATERAL CONTROL RESEARCH

30. NACA TECHNICAL NOTE No. 1404 —
COLLECTION OF TEST DATA FOR LATERAL CONTROL WITH FULL SPAN FLAPS
Fischel and Ivey

31. NASA NGR 17-002-072 —
SPOILERS FOR ROLL CONTROL OF LIGHT AIRPLANES
Roskan - Wentz - Kohlman
University of Kansas, Lawrence, Kansas

32. LIGHT AIRPLANE DESIGN
L. Pazmany
P.O. Box 10051, San Diego, CA

33. NACA Report No. 927 —
APPRECIATION AND PREDICTION OF FLYING QUALITIES
Phillips (1949)

34. A PRACTICAL GUIDE TO AIRPLANE PERFORMANCE AND DESIGN
Donald R. Crawford
Crawford Aviation
P.O. Box 1262
Torrance, CA 90505

35. WIND TUNNEL INVESTIGATION OF A FULL SCALE CANARD CONFIGURED (VARIEZE) GENERAL AVIATION AIRCRAFT
Long P. YIP, Paul F. Coy
NASA Langley Research Center
Hampton, VA 23665

36. EFFECTS OF RAIN AND BUGS ON FLIGHT BEHAVIOR OF TAIL FIRST AIRPLANES
Don Hewes
Sport Aviation - May, June, July 1983

37. STATIC LONGITUDINAL STABILITY OF "ENTE" AIRPLANES
NACA T.M. 612
Heinrich Georg Kiel

38. FOCKE-WULF FIG. A "ENTE" COMMERCIAL ꞧPLANE (German)
NACA Aircraft Circular No. 132

39. DESIGN CONSIDERATIONS FOR STALL/SPIN SAFETY
Burt Rutan
Sport Aviation September 1974

40. NACA Report 651 —
DOWNWASH AND WAKE BEHIND PLAIN AND FLAPPED AIRFOILS
Silverstein - Katzoff and Bullivant (1939)

WARNING—A WORD OF CAUTION

Flight, in and of itself, is not necessarily dangerous, however it is most unforgiving of errors, sloppiness, and misjudgements on the parts of the designer, manufacturer, and pilot. Whenever a man flies, he accepts the risk that he may be injured or even killed. It is each individual's decision to either accept or reject this risk in light of its potential hazards, challenges, and rewards. Flying can be, and is done safely every day of the year, by paying strict attention to the details involved.

This book is not intended as a do-it-yourself guide, but merely as a source of information to be used as a reference. If there is anything you don't understand, don't hesitate to ask your flight instructor, other pilots, and experts. Many of the aircraft featured in this volume are homebuilt, and as such, are governed by the FAA's Advisory Circular No. 20-27C—Certification And Operation Of Amateur-Built Aircraft. Canards handle differently than conventional aircraft, and appropriate skills are necessary to operate them. And, as always, remember to treat any aircraft with due respect.

www.ingramcontent.com/pod-product-compliance
Lightning Source LLC
Chambersburg PA
CBHW081503200326
41518CB00015B/2365